Lecture Notes in Computer S

Commenced Publication in 1973
Founding and Former Series Editors:
Gerhard Goos, Juris Hartmanis, and Jan van Leeu

Srinath Srinivasa   Vasudha Bhatnagar (Eds.)

Big Data Analytics

First International Conference, BDA 2012
New Delhi, India, December 24-26, 2012
Proceedings

 Springer

Volume Editors

Srinath Srinivasa
International Institute of Information Technology
26/C, Electronics City, Hosur Road, Bangalore 560100, India
E-mail: sri@iiitb.ac.in

Vasudha Bhatnagar
University of Delhi, Faculty of Mathematical Sciences
107, Department of Computer Science
Delhi 110007, India
E-mail: vbhatnagar@cs.du.ac.in

ISSN 0302-9743   e-ISSN 1611-3349
ISBN 978-3-642-35541-7   e-ISBN 978-3-642-35542-4
DOI 10.1007/978-3-642-35542-4
Springer Heidelberg Dordrecht London New York

Library of Congress Control Number: 2012953578

CR Subject Classification (1998): H.3, I.2, H.4, H.2.8, I.4, H.5

LNCS Sublibrary: SL 3 – Information Systems and Application, incl. Internet/Web
and HCI

Typesetting: Camera-ready by author, data conversion by Scientific Publishing Services, Chennai, India

Printed on acid-free paper

Springer is part of Springer Science+Business Media (www.springer.com)

Preface

The arrival of the so-called *Petabyte Age* has compelled the analytics community to pay serious attention to development of scalable algorithms for intelligent data analysis. In June 2008, *Wired* magazine featured a special section on "The Petabyte Age" and stated that "..our ability to capture, warehouse, and understand massive amounts of data is changing science, medicine, business, and technology." The recent explosion in social computing has added to the vastly growing amounts of data from which insights can be mined. The term "Big Data" is now emerging as a catch-all phrase to denote the vast amounts of data at a scale that requires a rethink of conventional notions of data management.

There is a saying among data researchers that "more data beats better algorithms." Big Data provide ample opportunities to discern hitherto inconceivable insights from data sets. This, however, comes with significant challenges in terms of both computational and storage expense, of the type never addressed before. Volume, velocity, and variability in Big Data repositories necessitate advancing analytics beyond operational reporting and dashboards. Early attempts to address the issue of scalability were handled by development of incremental data mining algorithms. Other traditional approaches to solve scalability problems included sampling, processing data in batches, and development of parallel algorithms. However, it did not take long to realize that all of these approaches, except perhaps parallelization, have limited utility.

The International Conference on Big Data Analytics (BDA 2012) was conceived against this backdrop, and is envisaged to provide a platform to expose researchers and practitioners to ground-breaking opportunities that arise during analysis and processing of massive volumes of distributed data stored across clusters of networked computers. The conference attracted a total of 42 papers, of which 37 were research track submissions. From these, five regular papers and five short papers were selected, leading to an acceptance rate of 27%.

Four tutorials were also selected and two tutorials were included in the proceedings. The first tutorial entitled "Scalable Analytics: Algorithms and Systems" addresses implementation of three popular machine learning algorithms in a Map-Reduce environment. The second tutorial, "Big-Data: Theoretical, Engineering and Analytics Perspectives," gives a bird's eye view of the Big Data landscape, including technology, funding, and the emerging focus areas. It also deliberates on the analytical and theoretical perspectives of the ecosystem.

The accepted research papers address several aspects of data analytics. These papers have been logically grouped into three broad sections: Data Analytics Applications, *Knowledge Discovery Through Information Extraction*, and *Data Models in Analytics*.

In the first section, Basil et al. compare several statistical machine learning techniques over electro-cardiogram (ECG) datasets. Based on this study, they make recommendations on features, sampling rate, and the choice of classifiers in a realistic setting. Yasir et al. present an approach for information require- ments elicitation (IRE), which is an interactive approach for building queries, by asking a user his/her information needs. Gupta et al. look at Big Data from the perspective of database management. They divide analytics over Big Data into two broad classes: data in rest and data in motion, and propose separate database solutions for both of them. Reddy et al. describe their efforts in im- parting practical education in the area of agriculture by means of a virtual lab. A virtual "crop lab" designed by the authors contains large amounts of practical data about crops that are indexed and summarized. The authors speculate on pedagogic methodologies necessary for imparting practical education using such crop data.

In the second section, Yasir et al. address the problem of schema summa- rization from relational databases. Schema summarization poses significant chal- lenges of semantically correlating different elements of the database schema. The authors propose a novel technique of looking into the database documentation for semantic cues to aid in schema summarization. Nambiar et al. present a faceted model for computing various aspects of a topic from social media datasets. Their approach is based on a model called the *colon classification scheme* that views social media data along five dimensions: Personality, Matter, Energy, Space, and Time. Gupta et al. address text segmentation problems and propose a technique called Analog Text Entailment that assigns an entailment score to extracted text segments from a body of text in the range [0,1], denoting the relative importance of the segment based on its constituent sentences. Kumar et al. study the price movements of gold and present a model explaining the price movements using three feature sets: macro-economic factors, investor fear features, and investor behavior features.

Finally, in the last section, Ryu et al. propose a method for dynamically generating classifiers to build an ensemble of classifiers for handling variances in streaming datasets. Mohanty and Sajith address the problem of eigenvalue com- putations for very large nonsymmetric matrix datasets and propose an I/O effi- cient algorithm for reduction of the matrix dataset into Hessenberg form, which is an important step in the eigenvalue computation. Gupta et al. address the problem of information security in organizational settings and propose a notion of "con- text honeypot" – a mechanism for luring suspected individuals with malicious in- tent into false areas of the dataset. They analyze the elements required for luring conditions and propose a mathematical model of luring. Kim et al. address recom- mender systems for smart TV contents, which are characterized by large sparse matrix datasets. They propose a variant of collaborative filtering for building effi- cient recommender systems. Kumar and Kumar address the problem of selecting optimal materialized views over high-dimensional datasets, and propose simulated annealing as a solution approach.

We would like to extend our gratitude to the supporting institutes: University of Delhi, University of Aizu, Indian Institute of Technology Delhi, and ACM Delhi-NCR Chapter. Thanks are also due to our sponsors: Microsoft Corporation, Hewlett Packard India, and IBM India Research Lab. And last but not the least, we extend our hearty thanks to all the Program, Organizing, and Steering Committee members, external reviewers, and student volunteers of BDA 2012.

December 2012   Srinath Srinivasa
                                                                 Vasudha Bhatnagar

Organization

BDA 2012 was organized by the Department of Computer Science, Univeristy of Delhi, India, in collaboration with University of Aizu, Japan.

Steering Committee

N. Vijyaditya	Ex-Director General (NIC), Government of India
Ajay Kumar	Dean (Research), University of Delhi, India
R.K. Arora	Ex-Professor, IIT Delhi, Delhi, India
Rattan Datta	Ex-Director, Indian Meteorological Department, Delhi, India
Pankaj Jalote	Director, IIIT Delhi, India
Jaijit Bhattacharya	Director, Government Affairs, HP India

Executive Committee

Conference Chair

S.K. Gupta	IIT, Delhi, India

Program Co-chairs

Srinath Srinivasa	IIIT, Banglore, India
Vasudha Bhatnagar	University of Delhi, India

Organizing Co-chairs

Naveen Kumar	University of Delhi, India
Neelima Gupta	University of Delhi, India

Publicity Chair

Vikram Goyal	IIIT, Delhi, India

Publications Chair

Subhash Bhalla	University of Aizu, Japan

Sponsorship Chair

DVLN Somayajulu	NIT Warangal, India

Local Organizing Committee

Ajay Gupta	University of Delhi, India
A.K. Saini	GGSIP University, Delhi, India
Rajiv Ranjan Singh	SLCE, University of Delhi, India
R.K. Agrawal	JNU, New Delhi, India
Sanjay Goel	JIIT, University, Noida, India
Vasudha Bhatnagar	University of Delhi, India
V.B. Aggarwal	Jagannath Institute of Management Sciences, New Delhi, India
Vikram Goyal	IIIT Delhi, India

Program Committee

V.S. Agneeswaran	I-Labs, Impetus, Bangalore, India
R.K. Agrawal	Jawaharlal Nehru University, New Delhi, India
Srikanta Bedathur	Indraprastha Institute of Information Technology Delhi, India
Subhash Bhalla	University of Aizu, Japan
Vasudha Bhatnagar	University of Delhi, India
Vikram Goyal	Indraprastha Institute of Information Technology Delhi, India
S.K. Gupta	Indian Institute of Technology, Delhi, India
S.C. Gupta	National Informatics Center, Delhi, India
Ajay Gupta	University of Delhi, India
Sharanjit Kaur	University of Delhi, India
Akhil Kumar	Pennsylvania State University, USA
Naveen Kumar	University of Delhi, India
C. Lakshminarayan	Hewlett-Packard Labs, USA
Yasuhiko Morimoto	Hiroshima University, Japan
Saikat Mukherjee	Siemens India
Peter Neubauer	Neo Technologies, Sweden
Sanket Patil	Siemens India
Jyoti Pawar	University of Goa, India
Lukas Pichl	International Christian University, Japan
Maya Ramanath	Indian Institute of Technology Delhi, India
B. Ravindran	Indian Institute of Technology Madras, India
Pollepalli Krishna Reddy	International Institute of Information Technology Hyderabad, India
S.H. Sengamedu	Komli Labs, India
Myra Spiliopoulou	University of Magdeburg, Germany
Srinath Srinivasa	International Institute of Information Technology Bangalore, India

Ashish Sureka   Indraprastha Institute of Information Technology,
                               Delhi, India
Shamik Sural   Indian Institute of Technology Kharagpur, India
Srikanta Tirthapura   University of Iowa, USA

Sponsors

University of Delhi, India
University of Aizu, Japan
IBM India Research Institute
Hewlett-Packard India
Indian Institute of Technology Delhi, India
ACM Delhi-NCR Chapter, India
Microsoft Corp.

Table of Contents

Data Models in Analytics

Scalable Analytics – Algorithms and Systems

Srinivasan H. Sengamedu

Komli Labs, Bangalore
shs@komli.com

Abstract. The amount of data collected is increasing and the time window to leverage this has been decreasing. To satisfy the twin requirements, both algorithms and systems have to keep pace. The goal of this tutorial is to provide an overview of the common problems, algorithms, and systems for handling large-scale analytics tasks.

1 Introduction

Inexpensive storage, sensors, and communication have lead to an explosion in data. There is also in increasing trend to arrive at data-driven decisions: heuristics for decision making are increasingly getting replaced by data-driven and predictive approaches and insights. Converting massive data to crisp insights and actioning on them requires appropriate algorithms and systems. In this tutorial, we provide an overview of scalable algorithms and systems which make this possible.

The tutorial consists of six parts. The first three parts – *Applications*, *ML Problems*, and *MapReduce Landscape* – introduce big data applications, basic ML problems and techniques and MR technologies. The next three parts — *Scalable ML algorithms*, *ML Algorithms in MapReduce* and *ML Systems* — are more advanced and cover how to make basic ML algorithms scalable, discuss implementation of ML algorithms in MR, and the trends in Big Data Systems.

2 Applications

Big Data Analytics is real, diverse, and has the potential to make big impact. Applications abound in several fields – user modeling in Internet advertising, comment spam detection and content recommendation, real-time document indexing in web search, bioinformatics, and real time video surveillance.

Large-scale advertising systems typically deal with several hundred million users. Each user has several hundred events – site visits, search queries, ad views, ad clicks, etc. The activities of the users should be modeled with as recent data as possible. The scale and latency make this a very challenging application.

Spam on the Internet affects several applications – email, comments, blogs, etc. The scale of spam is very large – 85% ($\approx$ 75 Billion emails per day. See http://www.senderbase.org/home/detail_spam_volume?displayed= lastmonth) of email messages are spam. Spam also should be detected as fast as possible – otherwise there is a risk of impacting user experience for long periods of time.

S. Srinivasa and V. Bhatnagar (Eds.): BDA 2012, LNCS 7678, pp. 1–7, 2012.

Table 1. Machine Learning Problem and Algorithms

Task	Technique	Scalable
Nearest Neighbor	*kd*- Trees	LSH, Semantic Hashing
Classification	SVM	VW, SGD
Clustering	*k*-means	Canopy Clustering
Topic Models	LDA	PLDA
Regression	Logistic	SGD
Ranking	GBDT	GBDT
Recommendation	Matrix Factorization	SGD
Association Rule	A priori	Min-closed

3   ML Problems

Machine Learning and Data Mining address a variety of problems. Table 1 lists some of the problems and one representative algorithm for solving them. (The selection of the "representative" algorithm is somewhat arbitrary!)

Some general problems with common implementations of representative algorithms are listed below.

1. As the number of data points becomes large, the algorithms are not scalable. It has been observed that the loading of training data dominates the training time and in most applications the training data does not fit into the main memory.
2. In several applications, the number of features is also very large. In such cases, techniques which are successful in low-dimensional applications (like *kd*-trees), perform poorly when the number of dimensions exceed 30.

4   MapReduce Landscape

MapReduce is currently the popular paradigm for dealing with Big Data. The landscape is very rich and complex. See Figure 1 for some components in the Hadoop ecosystem and how they are related. A brief description of the different components follows.

HDFS: Hadoop Distributed File System - storage, replication
MapReduce: Distributed processing, fault tolerance
HBASE: Fast read/write access
HCatalog: Metadata
Pig: Scripting
Hive: SQL
Oozie: Workflow, Scheduling
Zookeeper: Coordination
Kafka: Messaging, Data integration
Mahout: Machine Learning

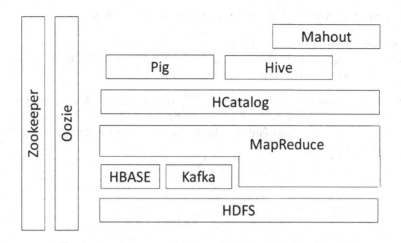

Fig. 1. Hadoop Ecosystem

5 Scalable ML Algorithms

We first review three promising ideas that lead to scalable machine learining – Random Projections, Stochastic Gradient Descent [3] which forms the core of several scalable algorithms, and the notion of MinClosed sequences. The last subsection lists some applications of these techniques.

5.1 Random Projections

Given a vector $\mathbf{x} \in \mathbb{R}^n$, we can define the projection of $\mathbf{x}$ on a random vector $\theta \in \mathbb{R}^n$ as

$$\theta^T \mathbf{x} + \gamma$$

A hash function $h(\cdot)$ based on the random projection is defined as

$$h^{\theta,\gamma}(\mathbf{x}) = \lfloor \frac{\theta^T \mathbf{x} + \gamma}{w} \rfloor$$

where w is the bin size. We can consider as the hash function as:

$$h : \mathbb{R}^n \to \{0, 1, \cdots, w - 1\}$$

Random projections have been used in a variety of ways. In Locality Sensitive Hashing, random projections are used to find nearest neighbors. In LSH, let the vectors $\mathbf{x}_1, \mathbf{x}_2, \cdots, \mathbf{x}_n$ constitute the database. We construct the hash table as follows.

Superhash function: This is defined as the concatenation of l hash functions.
$$g(\mathbf{x}) = h_1(\mathbf{x}) \cdot h_2(\mathbf{x}) \cdots h_l(\mathbf{x})$$
Hash Tables: Choose L such superhash functions $g_1(\mathbf{x}), g_2(\mathbf{x}), \cdots, g_L(\mathbf{x})$.

Given a query vector $\mathbf{q}$, we find the nearest neighbor as follows.

1. Calculate $g_1(\mathbf{q}), g_2(\mathbf{q}), \cdots, g_L(\mathbf{q})$.
2. For each $g_i(\mathbf{q})$, find the data points hashed to the same value by g_i. Stop as soon as a point in the required distance is found.

Similar ideas are used for *feature hashing* [13]. In this section, h is a hash function $h : \mathbb{N} \rightarrow \{1, \cdots, m\}$. Let ξ be a binary hash function. $\xi : \mathbb{N} \rightarrow \{\pm 1\}$. Given two feature vectors $\mathbf{x}$ and $\mathbf{x}'$, we define the hashed feature map and the corresponding inner product as

$$\phi_i^{h,\xi}(\mathbf{x}) = \sum_{j:h(j)=i} \xi(j)\mathbf{x}_j$$

and

$$< \mathbf{x}, \mathbf{x}' >_\phi = < \phi^{h,\xi}(\mathbf{x}), \phi^{h,\xi}(\mathbf{x}') >$$

It can be shown that the hash kernel is unbiased [13]. In other words,

$$\mathbf{E}_\phi[< \mathbf{x}, \mathbf{x}' >_\phi] = < \mathbf{x}, \mathbf{x}' >$$

5.2 SGD

Stochastic Gradient Descent is one of the simplest techniques for optimization and is surprisingly one of the most effective in large scale applications. Assume that we are given n training examples: $\{\mathbf{x}_i, y_i\}$ where $\mathbf{x}_i$ is the feature vector and y_i is the expected output. Assume that we are learning from a family $\mathcal{F}$ of functions with $\mathbf{w}$ being the parameter. The empirical loss of any specific function $f \in \mathcal{F}$ is

$$E_n(f) = \frac{1}{n} \sum_{i=1}^{n} \mathcal{L}(f(\mathbf{x}_i; \mathbf{w}), y_i)$$

The standard gradient descent update of $\mathbf{w}$ looks as below:

$$\mathbf{w}_{t+1} = \mathbf{w}_t - \gamma \frac{1}{n} \sum_{i=1}^{n} \nabla_w \mathcal{L}(f(\mathbf{x}_i; \mathbf{w}_t), y_i)$$

This is also called the batch update. The complexity of batch update is $O(n)$. SGD is a form of online update and the update depends only on the current example.

$$\mathbf{w}_{t+1} = \mathbf{w}_t - \gamma_t \nabla_w \mathcal{L}(f(\mathbf{x}_t; \mathbf{w}_t), y_t)$$

For convergence, the following conditions are needed: $\sum_t \gamma_t^2 < \infty$ and $\sum_t \gamma_t = \infty$.

5.3 MinClosed Sequences

Frequent pattern and sequence mining are very important problems and have several applications. A frequent sequence is a sequence whose support exceeds a pre-defined threshold. A closed sequence is the longest sequence for a given support. More formally, a sequence s is defined to be closed with respect to a set $\mathcal{D}$ if and only if there does not exist any sequence s' such that $s \preceq s'$ and $Support_{\mathcal{D}}(s) = Support_{\mathcal{D}}(s')$.

The notion of *MinClosed Sequence* [5] has additional advantage in the sense that it provide further reduction in the number of sequences to consider. Given a sequence database $\mathcal{D}$, a sequence s is defined to be *min-l-closed* if all the following hold.

1) $length(s) \geq l$

2) s is closed

3) There does not exist a closed subsequence $s' \preceq s$ with $length(s') \geq l$

The term *MinClosed* is used when the parameter l is obvious from the context.

The notion of *MinClosed sequences* is interesting because of the following result:

$$Support(\mathcal{C}_{min}) = Support(\mathcal{C}).$$

5.4 Applications

The techniques discussed in the above subsections have been used in several innovative applications: finding nearest neighbors through machine learning [10], identifying spam in comments using the notion of min-closed sequences [5], and inferring user interests [2] – to name a few.

6 ML Algorithms on MapReduce

In the framework of PAC Learning, Statistical Query Model (SQM) refers to the class of algorithms that use statistical properties of the data set instead of individual examples. [6] shows that any algorithm fitting SQM can be converted to a learning algorithm that is noise-tolerant. In addition to this interesting theoretical property, [4] shows that algorithms fitting SQM can be expressed in a "summation form" that is amenable to MR framework.

Since SGD has proved as an effective training technique for several objective functions, let us look at MapReduce implementation of it. SGD, being iterative, also highlights the challenges of MR for ML applications. It is conceptually simple to split do the gradient computation of examples in the Map tasks and perform the aggregation and update in the Reduce task. (SGD framework simplifies this to a great extent as the updates do not need to involve all training examples.) Even this simple approach is fraught with the following practical difficulties: (1) The reducer has to wait for all map tasks to complete leading to synchronization issues. (2) Each iteration is a separate MapReduce task and incurs severe system overheads.

Ensemble methods come in handy. In case of Ensemble Methods, we train n classifiers $C_1, C_2, \cdots, C_n$ instead of one. The classifiers are largely independent but can be correlated because of overlap in training examples. The final classification output is a function of the output of these classifiers. A simple technique is *Majority Voting* in which the class which is predicted by most classifiers is the final classification.

This simple example demonstrates some key limitations of MapReduce framework for ML algorithms: (1) Synchronization (2) System Overheads. So alternative frameworks or enhancements have been proposed. A simple example is that of shared memory system. It can be shown that in a shared memory environment, SGD cam be implemented in a *lockfree* fashion [11]. The primary risk of such an approach is data overwrites by multiple processes. The paper shows that the overwrites are rare when the data access is *sparse* and have negligible impact on accuracy.

[12] also addresses the issues of synchronization in the context of Topic Models (more specifically, Latent Dirichlet Allocation). The proposed solution has two aspects: approximate update and blackboard data structure for global updates.

7   ML Systems

As a case study, we can consider the the ML platform at Twitter [8] from both the system and ML algorithm aspects. ML functions are expressed primarily using Pig with additional UDFs and not in Java. The learner is embedded inside Pig storage function (`store` with appropriate UDFs). Scalable ML is supported through SGD and Ensemble methods.

MapReduce framework, while being widely available, is inadequate for machine learning algorithms. See Section 6. GraphLab [9] is a more flexible framework for expressing ML-type flows. The Data Model of Graph Lab consists of a *data graph* and *shared data table*. The computation model consists of *update function* specifying local computation and *sync mechanism* for global aggregation. In addition *consistency models* and *update scheduling* are supported.

[1] proposes a simpler *AllReduce* framework for scalable learning of linear models.

[7] argues makes a case for not moving away from MapReduce but using only MapReduce for ML algorithms.

8   Conclusion

Scalable Machine Learning techniques and Scalable Machine Learning Systems are emerging research areas. While there are a few dominant trends like Random Projections and Stochastic Gradient Descent, much of the field is wide open and provides an opportunity for research in algorithms and systems.

References

1. Agarwal, A., Chapelle, O., Dudik, M., Langford, J.: A reliable effective terascale linear learning system (2012), http://arxiv.org/abs/1110.4198
2. Ahmed, A., Low, Y., Aly, M., Josifovski, V., Smola, A.J.: Scalable distributed inference of dynamic user interests for behavioral targeting. In: KDD (2011)
3. Bottou, L., Bousquet, O.: The tradeoffs of large scale learning. In: Advances in Neural Information Processing Systems (2008)

4. Chu, C.-T., Kim, S.K., Lin, Y.-A., Yu, Y., Bradski, G.R., Ng, A.Y., Olukotun, K.: Map-Reduce for machine learning on multicore. In: NIPS (2006)
5. Kant, R., Sengamedu, S.H., Kumar, K.: Comment spam detection by sequence mining. In: WSDM (2011)
6. Kearns, M.: Efficient noise-tolerant learning from statistical queries. JACM (1998)
7. Lin, J.: MapReduce is good enough? if all you have is a hammer, throw away everything that's not a nail! (2012), http://arxiv.org/abs/1209.2191
8. Lin, J., Kolcz, A.: Large-scale machine learning at Twitter. In: SIGMOD (2012)
9. Low, Y., Gonzalez, J., Kyrola, A., Bickson, D., Guestrin, C., Hellerstein, J.: GraphLab: a new framework for parallel machine learning. In: UAI (2010)
10. Nair, V., Mahajan, D., Sellamanickam, S.: A unified approach to learning task-specific bit vector representations for fast nearest neighbor search. In: WWW (2012)
11. Niu, F., Recht, B., Re, C., Wright, S.J.: HOGWILD!: A Lock-Free Approach to Parallelizing Stochastic Gradient Descent. In: NIPS (2011)
12. Smola, A., Narayanamurthy, S.: An architecture for parallel topic models. In: VLDB (2010)
13. Weinberger, K., Dasgupta, A., Langford, J., Smola, A., Attenberg, J.: Feature hashing for large scale multitask learning. In: ICML (2009)

Big-Data – Theoretical, Engineering and Analytics Perspective

Vijay Srinivas Agneeswaran

Innovation Labs, Impetus Infotech (India) Pvt Ltd., Bangalore, India
`vijay.sa@impetus.co.in`

Abstract. The advent of social networks, increasing speed of computer networks, the increasing processing power (through multi-cores) has given enterprise and end users the ability to exploit big-data. The focus of this tutorial is to explore some of the fundamental trends that led to the Big-Data hype (reality) as well as explain the analytics, engineering and theoretical trends in this space.

1   Introduction

Google's seminal paper on Map-Reduce [(Ghemawat, 2008)] was the trigger that led to lot of developments in the Big-Data space. While not a fundamental paper in terms of technology (the map-reduce paradigm was known in the parallel programming literature), it along with Apache Hadoop (the open source implementation of the MR paradigm) enabled end users (not just scientists) to process large data sets on a cluster of nodes – a usability paradigm shift. Hadoop which comprises the MR implementation along with the Hadoop Distributed File System (HDFS) has now become the de-facto standard for data processing, with lot of Industrial game-changers having their own Hadoop cluster installations (including Disney, Sears, Walmart, AT&T, etc).

The focus of this tutorial talk is to give a bird's eye view of the Big-Data landscape, including the technology, funding and the emerging focus areas as well as the analytical and theoretical perspective of the ecosystem. The four broad pieces of the Hadoop puzzle include systems and infrastructure management (done by core Hadoop companies such as Hortonworks, Cloudera), data management (NoSQL databases), data visualization (Jaspersoft, Microstrategy with their Hadoop connectors) and the applications on top of the above three pieces. In terms of funding, the data visualization and analytics companies have garnered the most funds, along with the Hadoop provisioning/engineering companies. The emerging areas of the ecosystem include the video analytics, ad placements and the software defined networks (SDNs). The emerging focus of Big-Data analytics is to make traditional techniques such as market basket analysis, scale and work on large data sets – the approach of SAS and other traditional vendors to build Hadoop connectors. The other emerging approach for analytics focuses on new algorithms including machine learning and data mining techniques for solving complex analytical problems including those in video and real-time analytics.

S. Srinivasa and V. Bhatnagar (Eds.): BDA 2012, LNCS 7678, pp. 8–15, 2012.

The Big-Data ecosystem has also seen the emergence of the NoSQL distributed databases such as the Amazon's Dynamo (Klophaus, 2010)(Giuseppe DeCandia, 2007), Cassandra (Malik, 2010), MongoDB (Eelco Plugge, 2010). These emerged mainly due to the limitations (in terms of fault-tolerance and performance) of the HDFS. Some of the NoSQL DBs including Dynamo and Riak (Klophaus, 2010) are key-value stores, while MongoDB and CouchDB (J. Chris Anderson, 2010) are document stores. The third category is the columnar databases such as BigTable (Fay Chang, 2006) and Cassandra, with the last category being the graph databases such as Neo4j. The tutorial gives a new theoretical perspective of the NoSQL databases – named as the NLC-PAC classification. It refers to the design choices made by the NoSQL databases – under normal operations, the trade-off is between latency and consistency, while under a partition, the trade-off is availability or consistency. It must be noted that the partition bit is in line with the CAP theorem (Lynch, 2012), while the NLC-PAC is inspired by (Abadi, 2012)).

The rest of the paper is organized as follows: Section 2 explains the theoretical foundations of the NoSQL databases including the CAP theorem and the need to look beyond the CAP theorem to include consistency VS latency trade-offs during normal operation of distributed systems. Section 3 explains the current trends in the Big-data space, including the funding patterns, the different pieces of the Hadoop puzzle and the three broad R&D focus areas and the key players (startups to watch out for and the established ones) in each.

2   Theoretical Perspective of the NoSQL Databases

2.1   CAP Theorem and Its Origin

The CAP theorem was given by Eric Brewer, one of the founders of Inktomi and now with Google, in 2001 in the keynote of Principles of Distributed Computing (PODC), the ACM flagship conference in the area.

Brewer says in the CAP theorem that out of the three essential things Consistency, Availability and Partitions, only any two is achievable in a distributed system at any point of time. This implies that in the face of partitions, there is a trade-off between availability and consistency and only one is achievable - lot of systems trade-off consistency for improved availability, such as Amazon Dynamo and Cassandra, while others sacrifice availability for consistency, as in the Google Chubby lock system, which makes best-effort at availability in the presence of partitions, but is a strongly consistent system .

It must importantly be noted that the CAP theorem is not new - it is a special case of the general impossibility results well known in distributed systems theory . The impossibility result states that it is impossible to achieve fault-tolerant agreement in a purely asynchronous distributed system. A purely synchronous distributed system is one in which

1.   All nodes in the system have a local clock for time-keeping and all clocks are synchronized with each other globally.
2.   Message delivery is bounded - this means that the time for delivery of a message cannot exceed a well-known threshold value.

3. All processors are in synchronization - this means that if a processor takes n steps, all others are guaranteed to take at least 1 step.

Fault-tolerant agreement means that nodes in the system agree on a certain value, even in the presence of arbitrary faults. Why are we saying this is a generalization of the CAP theorem? Termination (in consensus terms, this means that all nodes eventually output a value) is a liveness property of the distributed system, while the agreement or consensus (which implies that every node must output the same value) is a safety property. The general impossibility result implies that it is impossible to achieve both safety and liveness in an asynchronous distributed system.

The CAP theorem is a specific case of the general theorem, with availability as the liveness property and consistency as the safety property and looking at only partition among the failure modes - there are other forms of failures equally important and possibly occurring more often such as fail-stop failures or Byzantine failures. Partition implies the system becomes asynchronous and one cannot achieve fault-tolerant consistency. The CAP theorem can be reworded to say that in a distributed system with communication failures (due to possibly partitions) implying possibility of message failures/arbitrary delays, it is impossible for a web service to implement atomic read/write shared memory that guarantees a response to every request - impossible for a protocol that implements read/write register to guarantee both safety (consistency) and availability (liveness) in a partition prone distributed system.

2.2   Beyond the CAP Theorem

Just to recap what we have described in the previous section, we explained the CAP theorem and said it is special case of a more generic formulation – that both liveness and safety cannot be both achieved at the same time in an asynchronous distributed system. We also noted that some NoSQL systems trade-off consistency for availability, as per the CAP theorem – including the Amazon Dynamo, Cassandra and Riak. They achieve a weaker form of consistency known as eventual consistency – updates are propagated to all replicas asynchronously, without guarantees on the order of updates across replicas and when they will be applied. Contrastingly, strict consistency would have required updates to be ordered (say by using a master to decide the order or by using a global time stamp) the same way across replicas or the use of Paxos kind of protocol .

While the CAP theorem is important and talks about important design trade-offs, it must be noted that it talks about what happens or what should be the design trade-off under a network partition. Network partition itself is a rare event, as explained by Stonebraker among others and is dependent on details of system implementation such as what kind of redundancy exists in switches, in nodes, in network routes? What kind of replication is done (is it over a wide area)? How carefully have network parameters been tuned? One can go on and on…. The CAP theorem DOES NOT talk about the normal operation of a distributed system when there are no partitions. What kind of trade-offs exist during the normal operation of a distributed system? To understand this question, we need to see beyond the CAP theorem. This has been observed by other researchers as well such as Daniel Abadi. In fact, the paper by Professor Abadi was the inspiration behind this formulation.

There is a view that most NoSQL systems sacrifice consistency because of the CAP theorem. This is not completely true. CAP theorem does not imply that consistency (along with availability) cannot be achieved under normal operations of a distributed system. In fact, CAP theorem allows for full ACID properties (along with availability) to be implemented under normal operation of a distributed system – only in the (rare) event of a network partition, the trade-off between consistency and availability kicks in. However, a very different trade-off exists under normal operation of a distributed system – it is related to latency. Thus, the important trade-off during the normal operation of a distributed system is related to consistency VS latency. Latency forces programmers to prefer local copies even in the absence of partitions (Ramakrishnan, 2012) – an important observation, which leads us to the latency consistency trade-off.

Why does this trade-off even exist? What are its implications? We will attempt to answer the above questions. Consider a distributed system where the designer has to replicate the data to make it available in case the original data becomes unavailable. There are only 3 choices in case of replicated data:

1. Data updates are sent to all replicas either synchronously or asynchronously.
   1.A.Asynchronous case: replica divergence (order of updates different) – if there is no agreement protocol or a centralized node – preprocessing node.
   1.B.Synchronous case: latency is high – needs a 2 Phase commit (2PC) kind of protocol.
2. Data updates sent to a data-item specific node – master for this data-item.
   1.A.Synchronous – involves latency
   1.B.Asynchronous – could be inconsistent if reads are from all nodes & consistent if reads are only from master.
   1.C.Quorum protocols – updates sent to W nodes, reads from any of R nodes, R+W>N for N nodes in the system for consistency reads.
3. Data updates sent to arbitrary location first – master not always the same node.

All the NoSQL systems make the decisions based on the above – for example, Amazon Dynamo, Riak and Cassandra have a combination of 2.C and 3, which means that the master for a data item can be different at different points in time and they implement a quorum protocol to decide which replica is to be updated and which replica is to be read from to be a weak or a strong read. One of the students in the class asked an interesting question, which may come your mind too – What is the need for a there a master when you have a decentralized quorum protocol? The quorum protocol is more for deciding who (which replica) are the members of the write set, who are the members of the read set and whether R+W should be >N (for strong reads) or <= N (for weak reads), while the master is the one who is responsible for sending the updates to the write set and deciding on order of updates, if required.

Another important NoSQL system, the PNUTS system from Yahoo (Brian F. C., 2008) chooses 2B – inconsistent or weak reads at reduced latency cost. The PNUTS system serves as the basis for a lot of Yahoo applications, starting from Weather to email to answers. Thus, availability at low latency is the most important factor – which is why they trade-off consistency for latency under normal operation. In the event of a partition, the PNUTS system makes the data item unavailable for updates, trading-off availability.

2.3   Categorizing NoSQL Databases

Just to recap what we talked about in the previous section, we explained the importance of the consistency-latency trade-off under normal operation of a distributed system, which is beyond the CAP theorem – which only talks about partitions and the consequent availability-consistency trade-off.

We give a new formulation of NoSQL databases, inspired by Prof. Abadi's PACELC conjecture. The formulation is itself is quite simple – can be depicted as NLC-PAC – what is the trade-off during normal operations of a distributed system (Latency VS consistency) and what is the trade-off under partitions (Availability-Consistency)?

Lot of NoSQL systems that relax consistency under partitions also relax consistency under normal operations and hence fall into the *NL-PA* category. This includes Basically Available Soft State Eventually Consistent (BASE) systems such as Amazon Dynamo, Riak and Cassandra. It must be noted that they all implement some form of quorum protocol to realize consistency. Just to recollect, quorum protocols have R readers, W writers, with R+W being set to be greater than N (for N nodes in the system) for reading the last written value. Weak reads are supported by setting R+W to be less than or equal to N.

The full Atomicity Consistency Isolation Durability (ACID) systems (which are similar to the traditional database systems) in the NoSQL space include BigTable from Google, Megastore and VoltDB (Stonebraker, 2012). These are systems for which consistency are needs are similar to sequential consistency and will pay any availability/latency cost to achieve that. These are classified as NC-PC systems in my conjecture.

The PNUTS system from Yahoo is the only system (that we know of) which falls into the NL-PC category. This means that under normal conditions, PNUTS will trade-off consistency (by allowing weak reads even from slaves, rather than only from the master for a data-item). Under a partition, instead of weakening consistency further, they make the data-item unavailable for updates (and hence trade-off availability – CP systems as per CAP categorization). This is because the master may get caught in a minority partition and slaves would have no way of reaching it – reads, like during normal operations can be from any slave (and are anyway weak).

MongoDB [9] falls under the NC-PA category – the system is strongly consistent during normal operations sacrificing latency, while under partitions application code must take care of the possible multiple masters and their conflict resolution – which in essence means consistency is left to the application developer to sort out under partitions.

3   Current Trends in Big-Data

The four broad pieces of the Hadoop puzzle include systems and infrastructure management (done by core Hadoop companies such as Hortonworks, Cloudera), data management (NoSQL databases – by companies such as 10gen, DataStax or the recent efforts such as LinkedIn's Voldemart DB (Sumbaly, R, 2012)), data visualization (Jaspersoft, Microstrategy or Tableu with their Hadoop connectors) and the applications on top of the above three pieces. In terms of funding, the data visualization and analytics companies have garnered the most funds, along with the Hadoop provisioning/engineering companies. The emerging areas of the ecosystem include the video analytics, ad placements and the software defined networks (SDNs).

The funding patterns are given below in pictorial form (this data has been culled out from various sources on the web). As can be observed from the figure, the main areas of funding are Hadoop provisioning (companies such as Hortonworks and MapR), analytics (companies like Parstream or Bloomreach) and visualization (companies such as Tableau). Other funding areas are video analytics (companies such as TubeMogul or Video Breakouts), Software Defined Networks (companies such as Arista are trying to offload the Reduce processing onto the Routers/Switches), data munging (the ability to convert data from different formats/sources into meaningful machine readable format – companies such as Metamarkets) and the NoSQL databases companies such as 10gen (MongoDB) and DataStax (Cassandra).

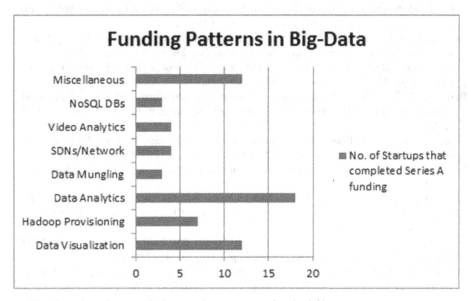

The three broad research themes that are emerging include:

1. *Storage, search and retrieval of Big-data*
Trends in this space include use of coding techniques such as Erasure Coding to optimize storage space, performing coding on graphics processing units, context sensitive search as well as Iceberg queries on Big-data. Companies such as Acunu

which is coming up with a custom storage for replacing HDFS by modifying the Linux kernel as well as companies such as Hadapt which allows SQL queries on distributed data sets are the interesting one to watch out for in this space.

2. *Analytics on Big-data*

Trends in this include real-time analytics by using Twitter's Storm integrated with Esper or other such query tools, video analytics and ad-placement analytics as well as incremental analysis using Google's Dremel . Interesting startups in this space include Parstream, Skytree, Palantir and Platfora. While some of the above start-ups are looking beyond Hadoop for analytics, Datameer is one company which is simplifying the use of Hadoop for Big-data analytics.

3. *Computations on Big-data*

Trends in this space include the moving beyond Map-Reduce paradigm using Google's Pregel and investigations of paradigms beyond the Map-Reduce for Big-data. Interesting startups in this space include Paradigm4, Black Sky, HPCC (which is exploring Hadoop alternatives) and YarcData (which has an alternative to the MR paradigm based on graph processing).

References

Abadi, D.J.: Consistency Tradeoffs in Modern Distributed Database System Design: CAP is Only Part of the Story. Computer 45(2), 37–42 (2012)

Baker, J.: Megastore: Providing Scalable, Highly Available Storage for Interactive Service. In: Conference on Innovative Data Systems Research, CIDR (2011)

He, B., Hsiao, H.-I.: Efficient Iceberg Query Evaluation Using Compressed Bitmap Index. IEEE Transactions on Data and Knowledge Engineering 24(9), 1570–1583 (2012)

Cooper, B.F., Ramakrishnan, R.: PNUTS: Yahoo!'s Hosted Data Serving Platform. Proceedings of VLDB Endowment, 1277–1288 (2008)

Burrows, M.: The Chubby Lock Service for Loosely-coupled Distributed Systems. In: ACM Symposium on Operating System Design and Implementation (OSDI), pp. 335–350. ACM (2007)

Huang, C., Simitci, H.: Erasure Coding in Windows Azure Storage. In: USENIX Conference on Annual Technical Conference (USENIX ATC 2012), p. 2. USENIX Association (2012)

Dirolf, M., Chodorow, K.: MongoDB: The Definitive Guide, 1st edn. O'Reilly Media, Inc. (2010)

Chang, F., Dean, J.: Bigtable: A Distributed Storage System for Structured Data. In: 7th USENIX Symposium on Operating Systems Design and Implementation, p. 15. USENIX Association, Berkeley (2006)

DeCandia, G., Hastorun, D.: Dynamo: Amazon's highly available key-value store. In: ACM Symposium on Operating Systems Principles, pp. 205–220. ACM (2007)

Malewicz, G., Matthew, H.: Pregel: A System for Large-Scale Graph Processing. In: SIGMOD International Conference on Management of Data (SIGMOD 2010), pp. 135–146. ACM, NY (2010)

Lindsay, B.S.: Single and Multi-Site Recovery Facilities. In: Poole, I.W. (ed.) Distributed Databases. Cambridge University Press (1980)

Lynch, N., Gilbert, S.: Brewer's conjecture and the feasibility of consistent, available, partition-tolerant web services. SIGACT News (June 2002)

Fischer, M.J., Lynch, N.A.: Impossibility of Distributed Consensus with one Faulty Process. Journal of the ACM 32(2), 374–382 (1985)

Malik, P., Lakshman, A.: Cassandra: A Decentralized Structured Storage System. SIGOPS Operating Systems Review 44(2), 35–40 (2010)

Meyer, M.: The Riak Handbook (2012), http://riakhandbook.com

Ramakrishnan, R.: CAP and Cloud Data Management. Computer 45(2), 43–49 (2012)

Sumbaly, R., Kreps, J.: Serving Large-scale Batch Computed Data with Project Voldemort. In: 10th USENIX Conference on File and Storage Technologies (FAST 2012), p. 18. USENIX Association, Berkeley (2012)

Al-Kiswany, S., Gharaibeh, A., Santos-Neto, E.: StoreGPU: Exploiting Graphics Processing Units to Accelerate Distributed Storage Systems. In: 17th International Symposium on High Performance Distributed Computing (HPDC 2008), pp. 165–174. ACM, NY (2008)

Melnik, S., Gubarev, A.: Dremel: Interactive Analysis of Web-Scale Datasets. Communications of the ACM 54(6), 114–123 (2011)

Stonebraker, M.: CACM Blog (2010), http://m.cacm.acm.org/blogs/blog-cacm/83396-errors-in-database-systems-eventual-consistency-and-the-cap-theorem/comments

Stonebraker, M.: Volt DB Blogs (2012), http://voltdb.com

Chandra, T.D., Griesemer, R.: Paxos Made Live: An Engineering Perspective. In: Twenty-Sixth Annual ACM Symposium on Principles of Distributed Computing (PODC 2007), pp. 398–407. ACM (2007)

A Comparison of Statistical Machine Learning Methods in Heartbeat Detection and Classification

Tony Basil[1], Bollepalli S. Chandra[1], and Choudur Lakshminarayan[2]

[1] Indian Institute of Technology, Hyderabad, India
{cs11m10,bschandra}@iith.ac.in
[2] HP Labs, USA
Choudur.Lakshminarayan@hp.com

Abstract. In health care, patients with heart problems require quick responsiveness in a clinical setting or in the operating theatre. Towards that end, automated classification of heartbeats is vital as some heartbeat irregularities are time consuming to detect. Therefore, analysis of electro-cardiogram (ECG) signals is an active area of research. The methods proposed in the literature depend on the structure of a heartbeat cycle. In this paper, we use interval and amplitude based features together with a few samples from the ECG signal as a feature vector. We studied a variety of classification algorithms focused especially on a type of arrhythmia known as the ventricular ectopic fibrillation (VEB). We compare the performance of the classifiers against algorithms proposed in the literature and make recommendations regarding features, sampling rate, and choice of the classifier to apply in a real-time clinical setting. The extensive study is based on the MIT-BIH arrhythmia database. Our main contribution is the evaluation of existing classifiers over a range sampling rates, recommendation of a detection methodology to employ in a practical setting, and extend the notion of a mixture of experts to a larger class of algorithms.

Keywords: Heart arrhythmia, ECG, Classification, Mixture of Experts.

1 Introduction

Heartbeat patterns may be identified from an ECG signal which is represented by a cardiac cycle consisting of the well known P-QRS-T waveforms. These patterns include normal beats and abnormalities such as premature ventricular contraction, and supra-ventricular contraction as seen in Fig. 1. Early detection of these abnormal beats is potentially life-saving as beat irregularities correspond to heart arrhythmias. The ECG signal of a normal beat has a distinctive structure consisting of five successive deflections in amplitude; known as the P, Q, R, S, and T waves as seen in Fig. 2. Features such as the time duration of the QRS complex and the R amplitude are useful in capturing the characteristics of the different types of beats. Existing detection methods use the time and amplitude dependent signatures of the heartbeat cycle to make a detection decision.

S. Srinivasa and V. Bhatnagar (Eds.): BDA 2012, LNCS 7678, pp. 16–25, 2012.

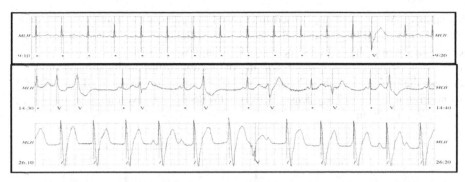

Fig. 1. Normal and pre-ventricular and supra ventricular beats observed in patients with cardiovascular diseases

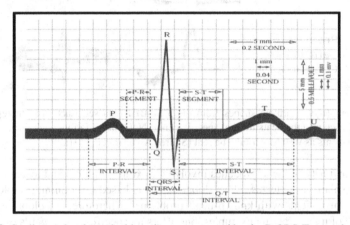

Fig. 2. Cardiac cycle of a typical heartbeat represented by the P-QRS-T wave form

Detection of irregularities in a heartbeat cycle is a fundamentally a challenging problem. In an ECG, the morphology of a beat type will not remain fixed, thus making it difficult to use standard template matching techniques. Fig. 3 depicts the heartbeat cycles of two patients (patient record numbers 119, 106 Source MIT-BIH). Clearly there is inter-patient and intra-patient variation. Therefore, any statistical algorithm should accommodate for these variations. So, adaptive methods are desirable, as they can quickly incorporate new patients' data. So a mixture of experts framework was used for ECG beat detection in [2]. In this approach, decision is based on a global and a local expert, which are self-organizing map (SOM) and learning vector quantization (VQ) classifiers that act on feature vectors extracted from each beat. Chazal et al [4] proposed classification scheme based on heartbeat morphology features, heartbeat interval features, and RR interval features and apply linear discriminant analysis (LDA) for classification. Ince et al [5] advanced a patient-dependent classification methodology for accurate classification of cardiac patterns. Wiens et al [10] proposed an active learning framework which reduces the dependence on patient-dependent labeled data for classification. Clearly heartbeat classification researches have focused exclusively on patient-wide and patient-specific ECG signatures. Alvarado et al [7] used time-based samplers, such as the Integrate-and-Fire (IF) model [8] to generate pulse based representations of the signal for heartbeat classification and showed that accurate ECG diagnosis is achievable. However, the focus of their paper was compression in conjunction with classification in order to address the problems such as band-width, power, and size encountered in real-time monitoring in wireless body sensor networks (WBSN), but nevertheless a solution to arrhythmia detection.

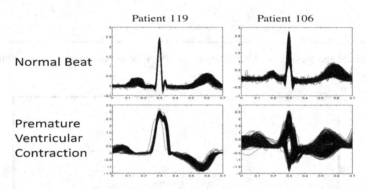

Fig. 3. Example of heartbeat shapes from the MIT-BIH data set. Each column represents a patient and each row the beats for that specific class. Note the variations in the beat morphology across patients as well as within a patient (Source Alvarado et al [7]).

As the majority of the approaches involve a mixture of global and local experts, we pause to explain these terms. The global expert is based on training samples from a population of patients with known heart problems. Therefore, the global expert has the benefit of many training examples, but offers no specificity in regards to a specific patient. To adapt to a single patient's unique characteristics, a local expert is trained on a small segment of a patient's annotated ECG data. A gating network dynamically weights the classification decisions of the global and local expert. Therefore, for typical beat morphologies, the global expert dominates, while the local expert tracks the more sensitive unique morphologies in rare beat patterns. Our approach incorporates a set of experts that train on the patient-wide database. When a new patient is presented, the expert that produces the smallest error is used in decision making. Our suite of classifiers consists of linear discriminant analysis (LDA), quadratic discriminant analysis (QDA), a mixture based on LDA and QDA, and the artificial neural networks based on the back propagation method. Results show that our approaches emulate and in many cases outperform adaptive mixtures in terms of typical classification criteria such as classifier sensitivity, specificity, and others.

The rest of the paper is organized as follows. In Section 2, we describe the data sources, description and pre-processing of the data used for analysis. In Section 3, we describe the various methods used, Section 4 discusses the results and recommendations, and finally Section 5, includes conclusions and next steps.

2   Data Description

We used the well-known MIT/Beth Israel Hospital (BIH) Arrhythmia Database available in the PhysioBank archives [1]. This database is composed of 2-channel ECG signals. For each patient, there is 30 minutes of ECG at a sampling rate of 360 Hz. We use the upper channel signal, which was acquired through a modified limb lead (MLL) with electrodes attached to the chest. The MIT/BIH Arrhythmia Database provides ECG signals with each beat labeled by an expert cardiologist. There are 20 types of beat annotations, as well as 22 types of non-beat annotations, including rhythm annotations. Testing protocol is carried out in compliance with American Association of Medical Instrumentation (AAMI) protocol [3]. The database contains 48 half-hour recordings obtained from 47 subjects. The signal is sampled at 360 Hz using 11 bit precision A/D converter over a ±5mV range. Adhering to the AAMI recommended standard [3], the four patients with paced beats were not considered for the training and testing. The remaining 44 patients were divided into two sets DS1 and DS2 where DS1 set was used for training the algorithms and DS2 set was used to evaluate the performance of various statistical classifiers. DS1, DS2 consist of the following patient records.

DS1 = {101, 106, 108, 109, 112, 114, 115, 116, 118, 119, 122, 124, 201, 203, 205, 207, 208, 209, 215, 220, 223, 230}

DS2 = {100, 103, 105, 111, 113, 117, 121, 123, 200, 202, 210, 212, 213, 214, 219, 221, 222, 228, 231, 232, 233, 234}

Paced beats = {102, 104, 107, 217}. Note that paced beats are excluded from analysis.

a.   ECG Filtering

Acquisition of ECG recordings involves detecting very low amplitude voltage in a noisy environment. Hence, we preprocessed the ECG signal to reduce the baseline wander and 60 Hz power line interference prior to applying the classification procedures for the detection of arrhythmia. To remove baseline wander, we passed the signal through median filters of window sizes 200ms and 600ms. This removes P-waves, QRS complexes and T-waves leaving behind the baseline wander. By subtracting the baseline wander from the original signal, we obtained the filtered signal. Furthermore, power line interference was removed by using a notch filter centered at 60Hz.

b.   Heartbeat Classes

The database has annotations for 20 different types of heartbeats. The annotation identifies the R-Peak for each heartbeat, where R-Peak represents the peak of QRS complex as seen in Fig. 2. In accordance with the AAMI standard [3], we grouped the heartbeat types into 5 classes. They are Normal and bundle branch block beats (N),

Supraventricular ectopic beats (SVEBs), Ventricular ectopic beats (VEBs), Fusion of normal and VEBs (F), and Unknown beats (Q). Table 1 shows the various heartbeat types listed in the MIT-BIH database. We performed binary classification task in compliance with AAMI [3] standard and consistent with studies reported in the literature, where Ventricular ectopic beats (VEB) are classified against the remaining heartbeat classes {N, S, F and Q}.

Table 1. MIT - BIH heartbeat group and the corresponding AAMI standard heartbeat class (Source Alvarado et al [7])

MIT BIH heartbeat group	AAMI heartbeat class	Beats
Normal beat Left bundle branch block beat Right bundle branch block beat Atrial escape beats Nodal (junctional) escape beat	N:Normal Beat	90631
Atrial premature beat Aberrated atrial premature beat Nodal(junctional) premature beat Supraventricular premature beat	S:Supraventricular ectopic beat	2781
Premature ventricular contraction Ventricular escape beat	V:Ventricular ectopic beat	7236
Fusion of ventricular and normal beat	F:Fusion beat	803
Paced beat Fusion of paced and normal beat Unclassified beat	Q:Unknown beat	8043

c.   Feature Extraction

We used a 13 dimensional feature vector, one for every heartbeat recorded in the 30 minute recording of each patient. The features consist of RR interval duration, the R-peak amplitude, and 5 samples each to the left and right of the R-peak, from the pre-processed ECG signal recordings down-sampled to 114 Hz. The 114 Hz sampling rate is in the vicinity of the average sample rate reported by Alvarado et al [7]. The RR interval features include the Pre-RR Interval and the Post-RR Interval. Pre-RR interval is calculated as the sample count between the current R-Peak and the preceding R-Peak, while Post-RR interval is calculated as the sample count between the current R-Peak and the next R-Peak. We settled for the 13 dimensional feature vector after it was found that having more sample values in the feature vector do not produce significant improvement in performance. It is noted that a lower sampling rate and smaller feature vector is desirable in real time monitoring applications as it translates to lower hardware complexity and power consumption.

3   Methods and Metrics

We used methods such as linear discriminant analysis (LDA), quadratic discriminant analysis (QDA), mixture of experts (ME), artificial neural networks (ANN) and ensemble networks. For an overview, refer to [11, 12, 13, 14]. In our implementation, the proposed mixture consists of two experts, LDA and QDA. The classification for each patient was performed using LDA and QDA and the winner between the two was chosen for each patient based on the F-Score, described later in this section. The LDA assumes that the covariance within the VEB and other generic class is the same, as opposed to the QDA which assumes an unequal covariance between the two classes. The ANN implementation consisted of network ensembles as they are known to exhibit superior performance in applications as widely reported in literature.

a.   Metrics

A variety of metrics are used in the realm of classification. Adhering to common practice in heartbeat classification, we used the metrics listed below. The classification results are reported in terms of accuracy (Acc), sensitivity (Se), positive predictive value (PPV), and false positive rate (FPR). They are defined as follows:

$$Accuracy\ (ACC) = \frac{TP+TN}{TP+TN+FP+FN}, Sensitivity(Se) = \frac{TP}{TP+FN},$$

$$Positive\ Predictive\ Value\ (PPV) = \frac{TP}{TP+FP}, and\ False\ Positive\ Rate(FPR) = \frac{FP}{TN+FP}$$

where TP (True Positive) is the number of heartbeats belonging to class $'i'$ that are correctly classified to class $'i'$; FN (False Negative) is the number of heartbeats belonging to class $'i'$ that are incorrectly classified to class $'j \neq i'$; FP (False Positive) is the number of beats of class $'j \neq i'$, that are incorrectly classified to class $'i'$; TN (True Negative) is the number of beats of class $'j \neq i'$ that are correctly classified to class $'j'$. In our model, VEB represents class $'i'$, and the set {N, F, Q, S} represents class $'j'$. An additional metric, the F score, was used in the mixture of experts model to compare the performance of LDA and QDA. F-Score is a common metric used in the field of Information Retrieval. Wiens et al [10] defined F-Score as

$$F - Score = \frac{2 * Se * PPV}{Se + PPV}$$

4   Results and Discussion

The classification scheme involved two classes consisting of feature vectors belonging to the VEB class, and the patterns from the remaining classes combined into one set {N, F, Q, S}. We have reported the results of the classification tasks in Table 2 and Table 3. Table 2 reports the gross results for LDA and QDA. Table 3 reports the gross results for ME (Mixture of experts) and ANN ensembles. Columns 2-6 contain the total number of heartbeats from each class; columns 7-14 report the gross classifier performance in terms of Acc (Accuracy), Se (Sensitivity), PPV (Positive predictive value) and FPR (False positive rate). Row 1 reports the gross result (Gross) for the 22 patient testing set DS2. Similarly, row 2 reports the gross result (Gross*) for the set of 11 patients overlapping with the testing set from Hu et al [2]. We used the aggregate TP, FN, FP, and TN to calculate the gross results for each classifier.

In the mixture of experts model (ME), the classification for each patient was performed using both LDA and QDA. For each patient, we chose the results for classifier (LDA or QDA) with higher F-Score, which was later used to calculate the gross results for the ME model. The gross statistics for the ANN ensemble classifier was calculated by taking the average of the results reported by the ANN ensemble.

Table 2. Gross classification results for LDA and QDA

Number of heartbeats						LDA				QDA			
	N	S	V	F	Q	Acc	Se	PPV	FPR	Acc	Se	PPV	FPR
Gross	44258	1837	3221	388	7	93.4	75.8	61.9	4.8	83.1	97	35.2	18.4
Gross*	23169	203	3174	388	2	94.2	75.8	75.3	3.3	83.6	97	41.6	18.2

Gross: Gross results for the 22 patient testing set

Gross*: Gross results for the 11 patients common to the testing set of Hu et al [2]

Table 3. Gross classification results for Mixture of Experts (ME) and ANN Ensemble

Number of heartbeats						Mixture of Experts				ANN Ensemble			
	N	S	V	F	Q	Acc	Se	PPV	FPR	Acc	Se	PPV	FPR
Gross	44258	1837	3221	388	7	95.4	92.2	63.4	4.3	96.9	79.7	74.6	1.9
Gross*	23169	203	3174	388	2	95.2	93	73.5	4.5	97.1	80.3	94.2	0.7

Gross: Gross results for the 22 patient testing set

Gross*: Gross results for the 11 patients common to the testing set of Hu et al [2]

The gross statistics (Gross) reported in Table 2 shows that LDA achieved higher Accuracy (93.4%), PPV (61.9%) and FPR (4.8%) while QDA achieved higher Sensitivity (97%). However, since QDA achieved a significantly lower PPV (35.2%) due to high false positives, LDA has clearly outperformed QDA. Comparing the ME model to the ANN ensemble in Table 3, the ANN ensemble clearly outperforms the ME model due to higher PPV. ME achieved high sensitivity (92.2%) and low PPV (63.4%), whereas ANN ensemble achieved a modest Sensitivity (79.7%) and PPV (74.6%). Thus after extensive testing, applying the many classifiers, the ANN stood out with the best performance, followed by mixture of experts model (ME), LDA and QDA. ANN ensemble achieved a gross accuracy (Acc) of 96.9% while LDA, QDA and mixture of experts (ME) achieved 93.4%, 83.1% and 95.4% respectively. ANN ensemble achieved sensitivity (Se) of 79.7%, while LDA, QDA and ME achieved sensitivity of 75.8%, 97%, and 92.2% respectively. However, ANN ensemble achieved the highest PPV among the four classifiers (74.6% for ANN, 61.9% for LDA, 35.2% for QDA and 63.4% for Mixture of experts). Therefore, the performance of the ANN ensemble has been found to be consistently superior to LDA, QDA and the ME.

In Table 4, we compare our methods with the state of the art models. The table shows the results reported by Chazal et al [4], Hu et al [2] and Alvarado et al [7] followed by the results obtained using our approach. Column 1 identifies the study, columns 2-5 represents the gross results for the 22 patient testing set. Columns 6-9 represents the gross results reported by the various studies for the set of 11 patients overlapping with the testing set from Hu et al [2]. Column 10 represents the sampling rate. The mixture of experts model (MoE) proposed by Hu et al [2] consists of a global expert and a local expert. The experts are weighted to make a decision. Hu et al (GE) represents the results obtained using the global expert, while Hu et al (MoE) represents the results obtained using the mixture of experts model described previously.

Table 4. Comparison of results with the state of the art

Methods	VEB Gross				VEB Gross*				Sample Rate
	Acc	Se	PPV	FPR	Acc	Se	PPV	FPR	
Chazal et al [4]	97.4	77.7	81.9	1.2	96.4	77.5	90.6	1.1	360
Hu et al(GE) [2] (Source Chazal et al [4])	-	-	-	-	75.3	69.6	34.6	16.8	180
Hu et al(MoE) [2] (Source Chazal et al [4])	-	-	-	-	93.6	78.9	76	3.2	180
Alvarado et al [7]	-	92.4	94.82	0.4	-	-	-	-	114
Proposed LDA	93.4	75.8	61.9	4.8	94.2	75.8	75.3	3.3	114
Proposed QDA	83.1	97	35.2	18.4	83.6	97	41.6	18.2	114
Proposed ME	95.4	92.2	63.4	4.3	95.2	93	73.5	4.5	114
Proposed ANN Ensemble	96.9	79.7	74.6	1.9	97.1	80.3	94.2	0.7	114

VEB Gross: Gross result achieved for the 22 patient testing set

VEB Gross*: Gross result achieved for the 11 patients common to the testing set of Hu et al [2]

We compared our results for the 22 patient testing set (VEB Gross) with the results reported by Chazal et al [4]. Sensitivity (Se) achieved by LDA, QDA, and ME are comparable to the sensitivity reported by Chazal et al [4]. However, LDA, QDA and ME achieved lower PPV. The performance of ANN ensemble(Acc equal to 96.9%, Se equal to 79.7%, PPV equal to 74.6% and FPR equal to 1.9%) is comparable to the results reported by Chazal et al [4] (Acc equal to 97.4%, Se equal to 77.7%, PPV equal to 81.9% and FPR equal to 1.2%). Note that while our sampling rate was 114 Hz, Chazal et al [4] sampled at 360 Hz. Also, we compared our results with the results reported by Alvarado et al [7] (Se equal to 92.4%, PPV equal to 94.82%, FPR equal to 0.4%). Note that Alvarado et al [7] reported an average sampling rate of 117 Hz, which is in within the range of our sampling rate. Our models were outperformed by the model proposed by Alvarado et al [7], which achieved higher sensitivity (Se) and positive predictive value (PPV). In a future paper, we propose to examine other types of embeddings to represent the ECG recordings to serve as a feature vector to compete against models proposed by Alvarado et al [7]. The approach by Alvarado et al [7] comes with many caveats. It may be noted that the proposal by Alvarado et al [7] using the integral and fire (IF) models derives its advantages if the analog signal is sampled as opposed to the digital signal. Hardware implementations for the IF are now not commercially available and are still in the incipient laboratory stages.

Columns 6-9 in Table 4 reports the gross results (VEB Gross*) achieved for the 11 patients overlapping with the testing set reported by Hu et al [2]. We have observed that ANN ensemble (Acc equal to 97.1%, Se equal to 80.3%, PPV equal to 94.2% and FPR equal to 0.7%) and our mixture of experts (ME) model (Acc equal to 95.2%, Se equal to 93%, PPV equal to 73.5% and FPR equal to 4.5%) outperforms Hu et al [2] with MoE (Acc equal to 93.6%, Se equal to 78.9%, PPV equal to 76% and FPR equal to 3.2%). The performance of proposed LDA model (Acc equal to 94.2%, Se equal to 75.8%, PPV equal to 75.3% and FPR equal to 3.3%) is comparable to the Mixture of experts model proposed by Hu et al [2]. For the same 11 patients, the results for ANN ensemble is comparable to the results reported by Chazal et al [4] (Acc of 96.4%, Se of 77.5%, PPV of 90.6% and FPR of 1.1%).

The results indicate that artificial neural networks are better suited for the detection of VEB type arrhythmia. It was observed that varying the learning rate and hidden layer nodes had minimal impact on the performance. Also, increasing the sampling rate to 180 Hz did not produce significant gain in performance. Hence a sampling rate of 114 Hz was found to provide enough discriminatory power for the classification task. In short, our approach emulated the performance of the state of the art models at a lower sampling rate and a set of simple features.

5   Conclusion and Next Steps

The main contribution of the paper is to review the state of the art in classification of heartbeats using ECG recordings. This is a comprehensive study that consisted of a suite of classifiers and variants there of applied to a binary classification task of detecting VEB type arrhythmias using the MIT-BIH patient archives. By performing extensive set of experiments over a range of sampling rates and over a range of tuning parameters specific to various classifiers, we are able to tabulate and compare the performances of individual classifiers. The practitioner based on domain knowledge and comfortable with tolerances relative to detection accuracy can choose an appropriate classifier based on our findings. Our investigation suggests that a simple set of morphological features together with time and amplitude features from the P-QRS-T waveform sampled at a 114 HZ and a well trained (offline) ensemble of neural networks can yield satisfactory results.

We intend to pursue this line of research to further enhance the classifier performance using inter-patient and intra-patient feature vectors. The chaotic behaviors of heart arrhythmias also provide the opportunity to explore these conditions by a dynamical systems and non-linear time series methods.

Acknowledgements. This work was supported by the Dept. of Information Technology (DIT), Govt. of India under the Cyber Physical Systems Innovation Project.

References

1. Goldberger, A.L., Amaral, L.A.N., Glass, L., Hausdorff, J.M., Ivanov, P.C., Mark, R.G., Mietus, J.E., Moody, G.B., Peng, C.-K., Stanley, H.E.: PhysioBank, PhysioToolkit, and PhysioNet: Components of a new research resource for complex physiologic signals. Circulation 101(23), e215–e220 (2000)
2. Hu, Y.H., Palreddy, S., Tompkins, W.J.: A patient adaptable ECG beat classifier using a mixture of experts approach. IEEE Trans. on Biomedical Engineering 44(9), 891–900 (1997)
3. Mark, R., Wallen, R.: AAMI-recommended practice: testing and reporting performance results of ventricular arrhythmia detection algorithms. Tech. Rep. AAMI ECAR (1987)
4. de Chazal, P., O'Dwyer, M., Reilly, R.: Automatic classification of heartbeats using ecg morphology and heartbeat interval features. IEEE Transactions on Biomedical Engineering 51(7), 1196–1206 (2004)
5. Ince, T., Kiranyaz, S., Gabbouj, M.: A generic and robust system for automated patient-specific classification of ecg signals. IEEE Transactions on Biomedical Engineering 56(5), 1415–1426 (2009)
6. Tan, P.N., Steinbach, M., Kumar, V.: Introduction to Data Mining. Addison-Wesley (2005)
7. Alvarado, A.S., Lakshminarayan, C., Principe, J.C.: Time-based Compression and Classification of Heartbeats. IEEE Transactions on Biomedical Engineering 99 (2012)
8. Feichtinger, H., Principe, J., Romero, J., Singh Alvarado, A., Velasco, G.: Approximate reconstruction.of bandlimited functions for the integrate and fire sampler. Advances in Computational Mathematics 36, 67–78 (2012)
9. Mark, R., Moody, G.: MIT-BIH Arrhythmia Database (May 1997),
   http://ecg.mit.edu/dbinfo.html
10. Wiens, J., Guttag, J.: Active learning applied to patient-adaptive heartbeat classification. In: Lafferty, J., Williams, C.K.I., Shawe-Taylor, J., Zemel, R., Culotta, A. (eds.) Advances in Neural Information Processing Systems, vol. 23, pp. 2442–2450 (2010)
11. Johnson, R.A., Wichern, D.W.: Applied Multivariate Statistical Analysis, 3rd edn. Prentice Hall, Englewood Cliffs (1992)
12. Duda, R.O., Hart, P.E.: Pattern Classification and Scene Analysis. John Wiley & Sons, New York (1973)
13. McLachlan, G.J.: Discriminant Analysis and Statistical Pattern Recognition. John Wiley & Sons, New York (1992)
14. Freeman, J.A., Skapura, D.M.: Neural Networks, Algorithms, Applications, and Programming Techniques. Computation and Neural systems Series. Addision Wesley, Reading (1991)

Enhanced Query-By-Object Approach for Information Requirement Elicitation in Large Databases

Ammar Yasir[1], Mittapally Kumara Swamy[1],
Polepalli Krishna Reddy[1], and Subhash Bhalla[2]

[1] Center for Data Engineering
International Institute of Information Technology-Hyderabad
Hyderabad - 500032, India
[2] Graduate School of Computer Science and Engineering
University of Aizu, Aizu-wakamatsu, Fukushima 965-8580, Japan
yasir@students.iiit.ac.in, kumaraswamy@research.iiit.ac.in,
pkreddy@iiit.ac.in, bhalla@u-aizu.ac.jp

Abstract. Information Requirement Elicitation (IRE) recommends a framework for developing interactive interfaces, which allow users to access database systems without having prior knowledge of a query language. An approach called 'Query-by-Object' (QBO) has been proposed in the literature for IRE by exploiting simple calculator like operations. However, the QBO approach was proposed by assuming that the underlying database is simple and contains few tables of small size. In this paper, we propose an enhanced QBO approach called Query-by-Topics (QBT), for designing calculator like user interfaces for large databases. We use methodologies for clustering database entities and discovering topical structures to represent objects at a higher level of abstraction. The QBO approach is then enhanced to allow users to query by topics (QBT). We developed a prototype system based on QBT and conducted experimental studies to show effectiveness of the proposed approach.

Keywords: User Interfaces, Information Systems, Information Requirement Elicitation, Query-by-Object.

1 Introduction

Databases are more useful, when users are able to extract information from the database with minimal efforts. Writing a structured query in XQuery or SQL is a challenging task for normal users. Research efforts are going on to design efficient user interfaces which simplify the process of database access. Information Requirement Elicitation [1] proposes an interactive framework for accessing information. In this framework, it was proposed that user interfaces should allow users to specify their information requirements by means of adaptive choice prompts.

In the literature, Query-By-Object (QBO) approach has been proposed to develop user interfaces for mobile devices [2], GIS systems [3] and e-learning

S. Srinivasa and V. Bhatnagar (Eds.): BDA 2012, LNCS 7678, pp. 26–41, 2012.

systems [4] based on IRE framework. The QBO approach provides a web-based interface for building a query using multiple user level steps. The main advantage of this approach is simplicity to express a query. The QBO approach uses a database to store the objects and entities. However, for databases with large number of tables and rows, the QBO approach does not scale well.

In this paper, we propose an improved QBO approach, Query-by-Topics (QBT), to design user interfaces based on IRE framework which works on large relational databases. In the proposed approach, we represent the objects at a higher level of abstraction by clustering database entities and representing each cluster as a topic. Similarly, we organize instances of a entity in groups based on values of a user selected attribute. The aim of this paper is not to propose an approach for detecting topical structures but rather how such an approach can provide applications in practical scenarios like information systems. Experiments were conducted at the system and user level on a real dataset using a QBT based prototype and the results obtained are encouraging.

The rest of the paper is organized as follows. In Section 2, we explain the related works. In Section 3, we briefly explain the QBO approach and discovering topical structures in a database. In Section 4, we present the proposed framework. In Section 5, we present experiments and analysis of the proposed approach. The last section contains summary and conclusions.

2   Related Work

'Information Requirement Elicitation' [1] framework allows users to build their queries in a series of steps. The result of each step is used to determine the user's intent. However, this approach does not allow gathering of individual query responses to build complex queries. The 'Query-By-Object' (QBO) [3] approach on the other hand provides a high-level user interface for IRE which allows users to build a query progressively while also allowing gathering of individual query responses. We explain the QBO approach in Section 3.1.

Jagadish et al. [5] identified five difficulties associated with usability of databases. One of the difficulty is 'Painful Relations' which describes the consequences of schema normalization. What users perceive as a single unit of information is disintegrated by the normalization process. Yu and Jagadish [6] had proposed the earliest algorithms for generating schema summary for XML schema. However, some of the assumptions made for XML schema summarization were not valid for relational databases. Yang et al. [7], addressed these limitations and proposed a principled approach for relational schema summarization. In [8], the authors presented an efficient framework for discovering topical structures in databases by clustering database entities.

Other related works which aim to help users in query construction are: form based interaction and keyword search in databases. In an input form based querying mechanism [9], users submit their information needs via a form and get the results. In [10], the authors tried to automate the process of construction of query forms. With the help of a limited number of forms, the system is able to express a wide range of queries. This helps in leveraging the restriction on expressiveness posed by form based querying mechanisms. *Keyword* searches in databases [11] allow users to query databases using a set of keywords. However, query result validation is a challenging aspect is such systems. In [12], the authors proposed a new paradigm for data interaction called *guided interaction*, which uses interaction to guide the users in query construction. The authors proposed that databases should enumerate all possible actions available to the user, allowing users to explore and discover knowledge effectively.

The QBO approach does not scale well for large databases. In this paper, we address this issue by providing abstractions for objects and instances to help users locate their objects of interest easily. The approach in [7,6] was aimed to provide database administrators and analysts with simple, easy-to-read schema, while the approach in [8] was concerned with the issue of integrating databases. In this paper, we extend the notion of topic detection or summary generation for usage in information systems. The *Guided Interaction* paradigm, proposed in [12] is similar to the IRE framework. However, in IRE framework and the proposed approach, we do not enumerate all possible actions to the users, instead we provide quality enumerations which can be explored further to guide the user in constructing queries.

3   Background

The proposed approach proposes enhancements to the Query-by-Object approach. We also use the notion of detecting topical structures in databases to represent objects at a higher level of abstraction. Hence, in this section we explain the Query-By-Object Approach (QBO) in detail and also describe the framework for discovering topical structures in databases.

3.1   Query-By-Object Approach

The Query-By-Object (QBO) approach was proposed in [3], based on the notion of IRE. In this approach, the user communicates with a database through a high level interface. The initial intent of the user is captured via selection of objects from an object menu. The user navigates to select granularity of these objects and operators to operate between the selected objects. The user's actions are kept track in a query-bag, visible to the user at all stages. Finally, a SQL equivalent query is formulated and is executed at DBMS server. In the IRE framework, intermediate queries could not be utilized further and hence, there is not much support for complex queries. In QBO, user is allowed to gather and combine query results. This is supported by closure property of the interface. It states that the result of an operation on objects leads to formation of another object. Hence, the results of a query can be used to answer an extended query. As the QBO interface involves multiple user level steps, non-technical users can easily understand and use the system for retrieving information from the databases. The developer protocol and user protocol (Figure 1) for the QBO approach is as follows:

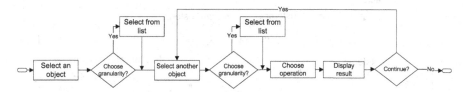

Fig. 1. QBO user protocol

Table 1. Operator Matrix for Example 1

	film	actor	fim_actor
film	U, I, C	R	R
actor	R	U, I, C	R
flim_actor	R	R	U, I, C

Table 2. QBO Developer and User Protocols

QBO Developer Protocol	QBO User Protocol
1. Store objects and entities in a RDBMS. 2. Define operators for each pair of objects. 3. Provide IRE based object selection, operation selection and support for closure property.	1. Select an object. 2. Select granularity of object. 3. Select another object. 4. Select the operator. 5. Display result. 6. If required, extend query by selecting another object.

Example 1. Consider an example where a developer builds a QBO based system which users will query.

System development based on QBO Developer Protocol: The following steps are taken by the developer:

- Database:
    - film - (*film_id*, film_name, film_rating)
    - actor - (*actor_id*, actor_name)
    - film_actor - (*fim_id, actor_id*, actor_rating);
- In this approach, the relations in the entity-relationship (ER) data model are considered as objects. Next, the developer defines pair wise operations between these objects (Table 1). Four types operators were proposed: UNION (U), INTERSECT (I), COMPLEMENT (C) and RELATE (R). The 'RE-LATE' operator has different connotations depending on the chosen objects it operates on. The pairwise operations are shown in Table 1.
- A web-based interface provides a list of objects, instances and operations user can select from. The system also allows user to combine query responses.

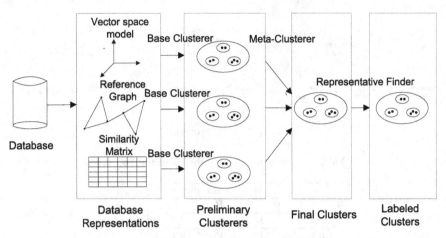

Fig. 2. The iDisc approach

Steps taken by user based on QBO User Protocol: Consider an example query that the user is interested in, *Find all actors who have co-worked with the actor 'Jack'*. Such a query when expressed with QBO can be expressed as: Find names of films actor 'Jack' has worked in, then find names of actors who worked in these films. User level steps are:

- Select object: actor
- Select granularity: actor-'Jack'
- Select another object: film
- Select operator: Relate
- Select another object: actor
- Select operator: Relate
- Display result

3.2   Discovering Topical Structures in Databases

Discovering topical structures in databases allows us to group semantically related tables in a single group, helping in identifying what users might perceive as a single unit of information in the database. Consider a database D, consisting of a set of tables $T = \{T_1, T_2...T_n\}$. Topical structure of a database D describes a partitioning, $C = \{C_1, C_2, ..C_k\}$ of tables in T such that the tables in the same partition have a semantic relationship and belong one subject area. In [8], the authors proposed *iDisc*, a system which discovers topical structure in a database by clustering tables into quality clusters.

The *iDisc* approach is described in Figure 2. The input to *iDisc* is a database D consisting of a set of tables T and returns a clustering C of the tables in T. In the *iDisc* approach, a database is first modeled by various representations namely *vector − based, graph − based* and *similarity − based*.

In the vector-based model, each table is represented as a document in bag-of-words model and a database is hence represented as set of documents. In the graph-based model, database is represented as an undirected graph. The nodes in the graph are the tables in the database(T). Two tables T_i and T_j share an edge if there exists a foreign key relationship between T_i and T_j. In the similarity-based representation a database D is represented as a $n \times n$ similarity matrix M where $n = |T|$ and $M[i,j]$ represents the similarity between tables T_i and tables T_j. The similarity between two tables is calculated by finding matching attributes based on a greedy-matching strategy [13]. The table similarity is then averaged out over the similarities of matching attributes.

In the next phase, clustering algorithms are implemented for each database representation model. The vector-based model and similarity-based model use hierarchical agglomerative clustering algorithm approach. A cluster quality metric defined to measure cluster quality. For the graph-based representation, shortest path betweenness and spectral graph partitioning techniques are used for partitioning the graph into connected components. Similar to other representations a cluster quality metric is used to measure quality of connected components. After clustering process ends, the base-clusterer for each representation selects the clustering with the highest quality score and preliminary clustering for each representation is discovered.

After identifying preliminary centerings, $iDisc$ uses a multi-level aggregation approach to aggregate results from each clustering using a voting scheme to generate final clusters. A clusterer boosting technique is also used in the aggregation approach by assigning weights to more accurate clustering representations. Later, representatives for each cluster is discovered using an importance metric based on centrality score of the tables in the graph-based representation. The output of $iDisc$ is a clustering of tables in the database, where each labeled clusters represents a topic.

4   Proposed Approach

The QBO approach has been discussed in Section 3.1. However, when the database size is large, the QBO approach does not scale well. We explain the issues using Example 2.

Example 2. Consider an example where a developer builds a QBO based system for a complex database $eSagu^{TR}$, described in section 5.1. For information requirement elicitation, following steps are taken by the system developer:

- Implement the $eSagu^{TR}$ system in a RDBMS, where each table would correspond to an object. The $eSagu$ database consists of 84 tables.
- Define operations between 84×84 object pairs.
- Provide a web-based interface providing a list of tables (84 tables) and instances (some tables containing more than 10^4 rows).

Use Case: Consider the scenario when a user is trying to query the $eSagu^{TR}$ database using a web-based interface designed using the developer's protocol. The user protocol would include:

- Select an object: a user would have to analyze a list of 84 objects and locate his object of interest.
- Select granularity or instance selection: Even if instance selection is based on attribute values, attributes can have large number of distinct values.
- Select operator: A user would have to have to grasp how each object would relate to other objects.

A complex database may contain large number of tables in the schema due to conceptual design or schema normalization. In such cases, it is difficult for the user to locate his information of interest. A naive solution, to organize objects alphabetically, may not be efficient. For example, in the eSagu database, there are 35 tables for various crop observations, cotton_observation, crossandara_observation and likewise 33 others. If a user wants to browse through all such observation tables he would need to know all the crop names. An organized list where crop observation tables are grouped together and then sorted alphabetically would be more intuitive for the user. Hence when the objects are more in number, there is a need to represent the objects at a higher level of abstraction. Similarly, there is a need for better organization when the object instances are more in number.

In general we are faced with the following problems for QBO developers and users:

- Large number of tables in the schema makes it harder for the user to locate his information of interest.
- With large number of instances in each table, selection of desired instance becomes difficult.
- With large number of tables, the number of pairwise operations between tables also increase. For n tables in the schema, in the worst case $n \times n$ operational pairs exist.

4.1 Basic Idea

In the proposed approach, we exploit the notion of detecting topical structures in databases to represent the schema at a higher level of abstraction. Identifying topical structures allows tables which are semantically correlated to be grouped together, which provides a better organization for options presented to the users. Secondly, instead of defining operations between each pair of tables, we can define operations between topics and within topics. Hence, the number of pairs for which operators have to be defined, can be reduced significantly. Similarly, to facilitate easier instance selection, we organize instances of an attribute into bins, providing a two-level hierarchy for instance selection. The developer protocol is modified to include steps required to generate the the abstract levels. Consequently, the user protocol is also modified for the proposed approach. The proposed approach has the following additional processes to QBO: Organizing objects into topics; Facilitating instance selection; Defining operators for the topical structure. We discuss each of these process in detail.

4.1.1. Organization into Topics

Discovering topical structures of objects stored in a RDBMS is a challenging task. While careful analysis of the database manually by a database domain expert would provide the most accurate result, in many cases it may be unfeasible to do so. In such cases, we use the *iDisc* approach described in section 3.2 for detecting topical structures in a database. The objective of this paper is not to compare various approaches for topic detection but rather focus on how such an approach has a practical use in information systems.

Given an input, a database containing a set of tables $T = (T_1, T_2, ..T_n)$, the *iDisc* framework generates a clustering $C = (C_1, C_2, ..C_k)$ of tables in the schema. along with the table representing the cluster centers (labels) $L = (L_1, L_2, ..L_k)$. C_i represents the set of tables belonging to the ith topic , where L_i represents the representative table (cluster center) and topic of the cluster C_i.

In QBO approach, the hierarchy of information organization was as follows:

$$\text{Tables} \rightarrow \text{Attributes} \rightarrow \text{Attribute Instances}$$

Given a clustering C and representative tables L, we make the following modification in the hierarchy of organization:

$$\text{Topics} \rightarrow \text{Tables} \rightarrow \text{Attributes} \rightarrow \text{Attribute Instances}$$

Each topic is represented by its representative-table. In other terms, we introduce topics and present the database tables belonging to a topic as its granularity. Hence, an object in QBT is a *topic* which has three levels of granularity (tables,attributes and attribute instances), in contrast to QBO which had only attributes and attribute instances as the two levels of granularity. Our approach is also in accordance to IRE framework. By providing topics, users can browse the database contents semantically, thus providing more intuitive options to the users.

4.1.2 Facilitating Instance Selection

For selecting instances of an object, selection based on an attribute values comes naturally to the user. Thus we first ask the user to select an attribute and then select its instances. However, in case the number of instances of an attribute are large, we need to an efficient organization of options. Here we have two problems in conflict as while we allow the user to drill down to his requirements in multiple steps, we may end up creating too many steps which is unfavorable for the user. We thus create a two-level hierarchy for attribute values such that there are not too many steps required for instance selection while providing a better organization. In the two-level hierarchy, we organize the attribute instances by grouping the attribute-instances into intervals. The first level represents the intervals and the second level represents the instances itself.

Considering instances of an attribute as a data distribution, we relate creating intervals to determine bins for creating histograms for a given data distribution. Methods for calculating number of bins given a data distribution are as follows:

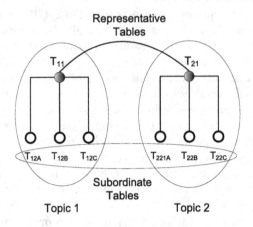

Fig. 3. Topical Structure for QBT

- Struge's formula: $k = \lceil log_2 n + 1 \rceil$
- Square root choice: $k = \sqrt{n}$
- Scott's choice (based on bin width): $h = \frac{3.5\sigma}{n^{\frac{1}{3}}}$, where h represents *Bin width*
- Freedman-Diaconis's choice: $h = 2 \times \frac{IQR(x)}{n^{\frac{1}{3}}}$, where $IQR = $ interquartile range.

4.1.3 Defining Operations
Next, we need to define operators which perform in case of QBT. Operators enable us to perform complex queries on databases involving one or more objects. The selected objects act as operands to the operators. We define two types of operator matrix:

i **Within-Topic Operator Matrix (WTS):** This matrix represents all possible operations within a topic. This includes operations between a topic's representative table with other tables belonging to the topic and between the tables in a same topic.

ii **Between-Topics Operator Matrix (BTS):** This matrix represents the possible operations between the representative tables of each topic. The diagonal elements represent the WTS matrix of the topics and other non-diagonal elements represent operations between two distinct topics.

By defining operational pairs between topics and within topic, we reduce the number of operation pairs for which operations need to be defined. The reduction in operational pairs depends on the topical structure identified for the database. Figure 3 shows an example of organization of tables into topical structure. A topic is represented by its representative-table and all tables belonging to a topic are called its subordinate tables. The first subscript represents the topic and second describes whether the table is a representative table or a subordinate table of the topic. Tables of each topic are further represented as a, b, and so on. Table 3 describes the Within Topic matrix for the first Topic (WT-I) and table 4 describes the Between Topic matrix (BT). The following scenarios come up in context of Figure 3,

Table 3. Within-Topic Matrix 1(WT-I)

t	T_{11}	T_{12a}	T_{12b}	T_{12c}
T_{11}	U,I,C	R	R	R
T_{12a}	R	U,I,C	R	R
T_{12b}	R	R	U,I,C	R
T_{12c}	R	R	R	U,I,C

Table 4. Between-Topic Matrix(BT)

t	T_{11}	T_{21}
T_{11}	[WT-I]	R
T_{21}	R	[WT-II]

Scenario 1. The two selected objects belong to the same topic. It has further three possibilities:

- *Both the tables are representative tables* $\{T_{11},T_{11}\}$: As there is only one representative table for each topic, this represents operations between same tables. The possible operations will be provided in Within-Topic operator matrix (WT-I[1,1]).

- *One of the table is representative-table and the other is a subordinate-table* $\{T_{11},T_{12a}\}$: This case represents a RELATE operation between the two tables. The operations will be defined in Within Topic operator matrix (WT-I[1,2]).

- *Both the tables are subordinate tables* $\{T_{12a},T_{12b}\}$: In this case, the two tables relate directly or through the representative-table of the corresponding topic. In this case, the operations are performed at a higher level (WT-I[2,3]).

Scenario 2. The two selected objects belong to different topics. It has three further possibilities:

- *Both the selected tables are representative-tables* $\{T_{11},T_{21}\}$: The possible operations will be defined in Between-Topics-I operator matrix (BT-I[1,2]).

- *One table is a representative-table and other is a subordinate table* $\{T_{11},T_{22a}\}$: In this case, the tables can be related at the higher level via the representative-tables of the two topics (BT-I[1,2]).

- *Both the tables are subordinate tables* ($\{T_{12a},T_{22a}\}$): Similar to the above case, the two tables can be related through their representative-tables. The possible operations are defined in Between-Topics-I matrix (BT[1,2]).

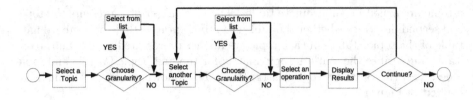

Fig. 4. QBT user protocol

4.2   QBT Protocols

In this section we describe the QBT developer protocol and QBT user protocol:

QBT Developer Protocol

- Store objects and entities in a database (RDBMS)
- Organize the tables in a schema based on the topic of tables, as described in Section 4.1.1.
- Create a framework to organize attribute instances into two-level hierarchy, as explained in Section 4.1.2.
- Define operations within each topic and between topics, described in Section 4.1.3.
- Provide a interface based on QBT, to allow object selection,instance selection and support closure property.

QBT User Protocol. The user protocol for QBT is described in Figure 4. The main options in the QBT are as follows.

- Select a topic
- Select granularity (a table, attribute and attribute values)
- Select another topic
- Select an operation
- Display result
- Extend query, if required

5   Experiments

5.1   Experimental Methodology

To analyze the effectiveness of the proposed approach, we conducted system-level experiments and also a usability study. System-level experiments consists of evaluating the reduction in the number of operational pairs from the QBO approach. The usability study consists task-analysis and ease of use survey on a real database using real users. For the usability study, we developed two prototypes, one based on the QBO approach and the other based on QBT approach. The interface for both the approach is almost similar, except that the QBO prototype does not group object by topics and does not not provide bins for instances. We first do a task analysis on the QBT prototype and QBO prototype to check whether the proposed approach is beneficial to the user. Since we give different tasks for different prototypes to the user, we cannot infer concrete results as different tasks are perceived differently by users. In order to overcome this bias, we ask the user to explore the database on their own and pose queries from their day-to-day requirements using both the prototypes. After the session, they fill out a questionnaire, rating the prototypes. It may not be the most efficient usability evaluation but it considerably reduces the bias from task analysis.

5.2 $eSagu^{TR}$ Database

For all experiments and analysis, we use a real database, $eSagu$. $eSagu^{TR}$ is a personalized agricultural advisory system. It provides a service through which expert agricultural advice is delivered to farmers for each of their farms regularly. In $eSagu^{TR}$, the agriculture scientist, rather than visiting the crop in person, delivers the expert advice by getting the crop status in the form of both digital photographs and the related information. The database consists of 84 tables containing mainly of farmer details, farm details, partner details, expert details, observation details and details of advice delivered for these observations.

5.3   Performance Analysis

We measure the effect of using topical structures at the system level by measuring the reduction factor (RF) for operational pairs. The reduction factor represents the number of operation pairs in the QBT approach as compared to the QBO approach (RF_{op}). If the number of operation pairs in QBT are OP_{qbt} and in case of QBO are OP_{qbo}, the reduction factor (RF_{op}) is defined as follows:

$$RF_{op} = 1 - \frac{OP_{qbt}}{OP_{qbo}} \tag{1}$$

We illustrate the metric by referring to Figure 3, where the total number of tables are 8. When tables have been divided into two topics, the number of operation pairs are are follows: two 4×4 WT matrix and one diagonal BT matrix (2 pairs). Hence OP_{qbt} is 34, while OP_{qbo} is 64 (8×8). The reduction factor for operation pairs is .46. For the eSagu database, after identifying topical structures, operational pairs were calculated for the *between topics matrix* (BT) and *within topics matrix* (WT). The reduction factor for operational pairs (RF_{op}) observed was .76.

5.4   Usability Study

Usability tests were conducted on four real users having computer experience but not skilled at SQL or query languages. The users belonged to the age group

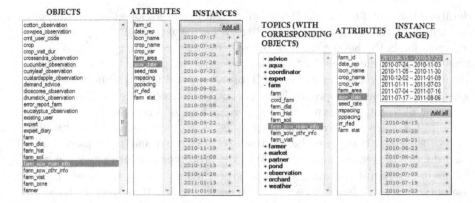

Fig. 5. QBO Approach Prototype

Fig. 6. QBT Approach Prototype(with Topic modeling and binning)

Table 4. Time taken and number of attempts made for each task

Task	Time taken in seconds (attempts taken)			
	User1	User2	User3	User4
1	21(1)	16(1)	41(2)	22(1)
2	18(2)	31(2)	30(1)	27(1)
3	170(3)	81(2)	79(1)	112(2)
4	17(1)	18(1)	22(1)	25(1)
5	25(1)	18(1)	41(2)	24(1)
6	140(2)	151(2)	110(2)	103(2)

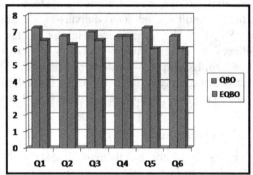

Fig. 7. Ratings from questionnaire

20-26 and were agriculture consultants at eSagu. The users were familiar with the database domain, mainly $eSagu^{TR}$ service and agricultural domain. and can validate the query results more comfortably. Users were then briefed about the QBT prototype for 15 minutes along with a quick demonstration of a sample query. Before the experiments, users were allowed a 5 minutes practice session to get themselves acquainted with the tool before starting the experiments. Using Retrospective testing, user's screen was recorded to monitor the navigation process. We performed two experiments: Task analysis and Use Survey [14].

Experiment 1, Task Analysis: Users were then asked to construct queries corresponding to the following six tasks on the QBT prototype and QBO prototype:

– Q1: Find the details of family members for the farmer *D.Laxama Reddy*.
– Q2: Find all the farmers owned by the farmer named *Polepally Thirumalreddy*.

- Q3: Find all the observations given to farmers from *Malkapur* village who grow *cotton* crops.
- Q4: Find the details of livestock belonging to the farmer *d.laxama reddy*.
- Q5: Find all the farmers belonging to the coordinator named *k .s narayana*.
- Q6: Find all the advices given to farmers from *Malkapur* village.

These three tasks (Q1,Q2,Q3) were performed on the QBT prototype and last three (Q4,Q5,Q6) were performed on the QBO prototype. The query Q1 is similar to query Q4, only difference on the part of the user would be to locate objects corresponding to the family details and livestock details. Similarly, Q2 and Q4 represent a simple join operation, differ only in terms of object involved. Q3 and Q6 represent a complex join involving 3 objects(4 tables from the database). Figure 5 and Figure 6 show the difference in prototypes for QBO approach and QBT approach. The tasks were performed in increasing order of complexity to track user's learning rate. The time taken(in seconds) and number of attempts taken to construct the query is described in Table 4. The experiments provided some interesting observations. The average time taken by users on first task was more than the fourth task. This may be because the user has already performed 3 queries on a similar interface and is now comfortable in using the interface for exploration or it may be that QBO prototype performs better. The average time taken for the second and fifth tasks are similar but the number of attempts made using the QBT prototype is less than that of QBO prototype, despite having used QBT prototype first. The third and sixth tasks took the most time as it involves a relatively complex join, with the average time taken for QBT prototype being less than QBO prototype. The difference is mainly due to location of objects quickly, during an attempt or relocating objects after a wrong attempt has been made.

Experiment 2, Use Survey: After the task evaluation, users were asked to explore the database on their own and pose various queries from their day-to-day requirements. Once the exploration session was finished, users were presented with another prototype, similar in design but based on earlier QBO approach. In the QBO based prototype, objects were not grouped and were presented by sorting them alphabetically in the objects menu. The users were again asked to explore the database using this prototype. After the users had explored the database using the two different prototypes they were asked to fill in a questionnaire based on a USE survey. The questionnaire asked the users to rate both the prototypes on a scale of 0 - 10 based on the following questions:

- Q1: The tool is easy to use.
- Q2: The tool sufficient for my information requirements
- Q3: The tool can be used with minimal efforts.
- Q4: The tool requires minimal training and can be used without written instructions.
- Q5: I can locate my information easily.
- Q6: The tool requires minimal steps to formulate a query.

In general we observe good results for most claims. For Q1, we achieved slightly higher rations for QBT than QBO, which shows that although the QBT approach consists of additional steps to the QBO, it does not deters the ease of use significantly. The claim Q2 received mixed reviews for both QBT and QBO prototype, which indicated that users demanded more operators to be provided to them so that their information needs can be elicited easily. For Q3 again QBT gets a higher rating than QBO which again emphasizes the ease of use of the QBT prototype. The claim Q4 got similar low ratings for QBO and QBT, indicating that users might require more briefings and demonstration to get comfortable in using the tool. The claim Q5 gets significantly higher ratings for QBT than QBO, which indicated the effectiveness of using topical structures and bins in large databases. For the claim Q6, we again get higher rating for QBO and QBT as users were able to operate between topics, ignoring the intermediate connections between the objects. In general, compared to the QBO prototype,the QBT prototype received better reviews for most the claims.

6 Summary and Conclusions

Accessing database requires user to be familiar with query languages. The QBO approach, based on IRE framework provides an interface where user progressively builds a query using multiple steps. This approaches works fine for small databases but cannot perform well for a database consisting of large number of tables and rows. In this paper, we propose Query-by-Topics, which provides enhancements over the existing QBO approach. We exploit topical structures in large databases to represent objects at a higher level of abstraction. We also organize instances of an object in a two-level hierarchy based on a user selected attribute. The advantages of this approach includes: user gets less navigational burden and the number of operations is reduced at the system level. The QBT prototype was implemented for a real database and experiments were conducted at the system level and user level to discuss the advantages. In future works, we we will work towards a more efficient approach for detecting topical structures.

References

1. Sun, J.: Information requirement elicitation in mobile commerce. Communications of the ACM 46(12) (2003)
2. Akiyama, T., Watanobe, Y.: An advanced search interface for mobile devices. In: HCCE 2012 (2012)
3. Rahman, S.A., Bhalla, S., Hashimoto, T.: Query-by-object interface for information requirement elicitation in m-commerce. International Journal of Human Computer Interaction 20(2) (2006)
4. Nemoto, K., Watanobe, Y.: An advanced search system for learning objects. In: HC 2010 (2010)
5. Jagadish, H.V., Chapman, A., Elkiss, A., Jayapandian, M., Li, Y., Nandi, A., Yu, C.: Making database systems usable. In: SIGMOD 2007 (2007)
6. Yu, C., Jagadish, H.V.: Querying complex structured databases. In: VLDB 2007. VLDB Endowment (2007)

7. Yang, X., Procopiuc, C.M., Srivastava, D.: Summarizing relational databases. Proc. VLDB Endow. (2009)
8. Wu, W., Reinwald, B., Sismanis, Y., Manjrekar, R.: Discovering topical structures of databases. In: SIGMOD 2008 (2008)
9. Choobineh, J., Mannino, M.V., Tseng, V.P.: A form-based approach for database analysis and design. Commun. ACM (1992)
10. Jayapandian, M., Jagadish, H.V.: Automating the design and construction of query forms. IEEE Trans. on Knowl. and Data Eng. (2009)
11. Qin, L., Yu, J.X., Chang, L.: Keyword search in databases: the power of rdbms. In: SIGMOD 2009 (2009)
12. Nandi, A., Jagadish, H.V.: Guided interaction: Rethinking the query-result paradigm. PVLDB (2011)
13. Rahm, E., Bernstein, P.A.: A survey of approaches to automatic schema matching. The VLDB Journal (2001)
14. Lund, A.: Measuring usability with the use questionnaire. Usability and User Experience Special Interest Group 8(2) (October 2001)
15. Borges, C.R., Macías, J.A.: Feasible database querying using a visual end-user approach. In: EICS 2010 (2010)

Cloud Computing and Big Data Analytics: What Is New from Databases Perspective?

Rajeev Gupta, Himanshu Gupta, and Mukesh Mohania

IBM Research, India
{grajeev,higupta8,mkmukesh}@in.ibm.com

Abstract. Many industries, such as telecom, health care, retail, pharmaceutical, financial services, etc., generate large amounts of data. Gaining critical business insights by querying and analyzing such massive amounts of data is becoming the need of the hour. The warehouses and solutions built around them are unable to provide reasonable response times in handling expanding data volumes. One can either perform analytics on big volume once in days or one can perform transactions on small amounts of data in seconds. With the new requirements, one needs to ensure the real-time or near real-time response for huge amount of data. In this paper we outline challenges in analyzing big data for both *data at rest* as well as *data in motion*. For big *data at rest* we describe two kinds of systems: (1) NoSQL systems for interactive data serving environments; and (2) systems for large scale analytics based on MapReduce paradigm, such as Hadoop, The NoSQL systems are designed to have a simpler key-value based data model having in-built *sharding*, hence, these work seamlessly in a distributed cloud based environment. In contrast, one can use Hadoop based systems to run long running decision support and analytical queries consuming and possible producing bulk data. For processing *data in motion*, we present use-cases and illustrative algorithms of data stream management system (DSMS). We also illustrate applications which can use these two kinds of systems to quickly process massive amount of data.

1 Introduction

Recent financial crisis has changed the way businesses think about their finances. Organizations are actively seeking simpler, lower cost and faster to market alternatives about everything. Clouds are cheap and allow businesses to off-load computing tasks while saving IT costs and resources. In cloud computing applications, data, platform, and other resources are provided to users as services delivered over the network. The cloud computing enables self-service with no or little vendor intervention. It provides a utility model of resources where businesses only pay for their usage. As these resources are shared across a large number of users, cost of computing is much lower compared to dedicated resource provisioning.

Many industries, such as telecom, health care, retail, pharmaceutical, financial services, etc., generate large amounts of data. For instance, in 2010, Facebook had 21 Peta Bytes of internal warehouse data with 12 TB new data added every day and 800

S. Srinivasa and V. Bhatnagar (Eds.): BDA 2012, LNCS 7678, pp. 42–61, 2012.

TB compressed data scanned daily [12]. These data have: large *volume*, an Indian Telecom company generates more than 1 Terabyte of call detail records (CDRs) daily; high *velocity*, twitter needs to handle 4.5 Terabytes of video uploads in real-time per day; wide *variety*, structured data (e.g., call detail records in a telecom company), semi-structured data (e.g., graph data), unstructured data (e.g., product reviews on twitter), which needs to be integrated together; and data to be integrated have different *veracity*, data needs to cleaned before it can be integrated. Gaining critical business insights by querying and analyzing such massive amounts of data is becoming the need of the hour.

Traditionally, data warehouses have been used to manage the large amount of data. The warehouses and solutions built around them are unable to provide reasonable response times in handling expanding data volumes. One can either perform analytics on big volume once in days or one can perform transactions on small amounts of data in seconds. With the new requirements, one needs to ensure the real-time or near real-time response for huge amount of data. The 4V's of big data – volume, velocity, variety and veracity—makes the data management and analytics challenging for the traditional data warehouses. Big data can be defined as data that *exceeds the processing capacity of conventional database systems*. It implies that the data count is too large, and/or data values change too fast, and/or it does not follow the rules of conventional database management systems (e.g., consistency). One requires new expertise in the areas of data management and systems management who understands how to model the data and prepare them for analysis, and understand the problem deeply enough to perform the analytics. As data is massive and/or fast changing we need comparatively many more CPU and memory resources, which are provided by distributed processors and storage in cloud settings. The aim of this paper is to outline the concepts and issues involved in new age data management, with suggestions for further readings to augment the contents of this paper. Here is the outline of this paper: in the next section, we describe the factors which are important for enterprises to have cloud based data analytics solutions. As shown in Figure 1, big data processing involves interactive processing and decision support processing of *data-at-rest* and real-time processing of *data-in-motion*, which are covered in Section 3, 4, and 5, respectively. For each data processing application, one may need to write custom code to carry out the required processing. To avoid writing custom code for data processing applications, various SQL like query languages have been developed. We discuss these languages in Section 6. Section 7 summarizes some enterprise applications which illustrate the issues one needs to consider for designing big data applications. Specifically, we consider applications in telecom, financial services, and sensor domains. We conclude by outlining various research challenges in data management over cloud in Section 8.

Figure 1 presents various components of big data processing story. This figure also mentions the section numbers corresponding to various components in this paper.

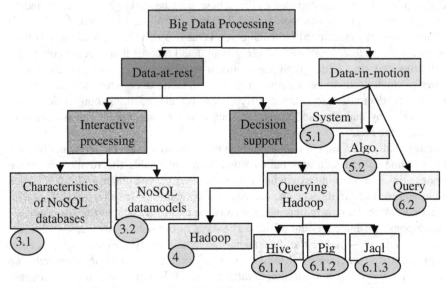

Fig. 1. Outline of the paper

2   Cloud Data Management

Cloud computing can be used for performing massive scale data analytics in a cost effective and scalable manner. In this section we discuss the interplay of cloud computing and data management, specifically, what are the benefits of cloud computing for data management; and the factors one should consider while moving from a dedicated data management infrastructure to a cloud based infrastructure.

2.1   Benefits of Cloud Computing

As mentioned earlier, a large volume of data is generated by many applications which cannot be managed by traditional relational database management systems. As organizations use larger and larger data warehouses for ever increasing data processing needs, the performance requirements continue to outpace the capabilities of the traditional approaches. The cloud based approach offers a means for meeting the performance and scalability requirements of the enterprise data management providing agility to the data management infrastructure. As with other cloud environments, data management in the cloud benefits end-users by offering a pay-as-you-go (or utility based) model and adaptable resource requirements that free up enterprises from the need to purchase additional hardware and to go through the extensive procurement process frequently. The data management, integration, and analytics can be offloaded to public and/or private clouds. By using public-cloud, enterprises can get processing power and infrastructure as needed, whereas with private-cloud enterprises can improve the utilization of existing infrastructure. By using cloud computing, enterprises can effectively handle the wide ranging database requirements with minimum effort, thus allowing them to focus on their core work rather than getting bogged down with the infrastructure. Despite all these benefits, decision to move from dedicated infrastructure to the cloud based data processing depends on several logistics and operational factors such as security, privacy, availability, etc.; which are discussed next.

2.2   Moving Data Management to Cloud

A data management system has various stages of data lifecycle such as data ingestion, ETL (extract-transform-load), data processing, data archival, and deletion. Before moving one or more stages of data lifecycle to the cloud, one has to consider the following factors:

1. **Availability Guarantees:** Each cloud computing provider can ensure a certain amount of availability guarantees. Transactional data processing requires quick real-time answers whereas for data warehouses long running queries are used to generate reports. Hence, one may not want to put its transactional data over cloud but may be ready to put the analytics infrastructure over the cloud.
2. **Reliability of Cloud Services:** Before offloading data management to cloud, enterprises want to ensure that the cloud provides required level of reliability for the data services. By creating multiple copies of application components the cloud can deliver the service with the required reliability of service.
3. **Security:** Data that is bound by strict privacy regulations, such as medical information covered by the Health Insurance Portability and Accountability Act (HIPAA), will require that users log in to be routed to their secure database server.
4. **Maintainability:** Database administration is a highly skilled activity which involves deciding how data should be organized, which indices and views should be maintained, etc. One needs to carefully evaluate whether all these maintenance operations can be performed over the cloud data.

Cloud has given enterprises the opportunity to fundamentally shift the way data is created, processed and shared. This approach has been shown to be superior in sustaining the performance and growth requirements of analytical applications and, combined with cloud computing, offers significant advantages [19]. In the next three sections, we present various data processing technologies which are used with cloud computing. We start with NoSQL in the next section.

3   NoSQL

The term NoSQL was first used in 1998 for a relational database that does not use SQL. It encompasses a number of techniques for processing massive data in distributed manner. Currently it is used for all the alternative data management technologies which are used for solving the problems for which relational databases are a bad fit. It enables efficient capture, storage, search, sharing, analytics, and visualization of the massive scale data. Main reason of using NoSQL databases is the scalability issues with relational databases. In general, relational databases are not designed for distributed horizontal scaling. There are two technologies which databases can employ for meeting scalability requirement: *replication* and *sharding*. In the *replication* technology, relational databases can be scaled using a master-slave architecture where reads can be performed at any of the replicated slave; whereas writes are performed at the master. Each write operation results in writes at all the slaves, imposing a limit to scalability. Further, even reads may need to be performed at master as previous write may not have been replicated at all the slave nodes. Although the situation can be improved by using multiple masters, but this can result in conflicts among masters whose resolution is very costly [28]. Partitioning (*sharding*) can also be used for scaling writes in relational databases, but applications are required to be made partition aware. Further, once partitioned, relations need to be handled at the application layer, defeating the very purpose of relational databases. NoSQL databases overcome these problems of horizontal scaling.

3.1   Characteristics of NoSQL Databases

An important difference between traditional databases and NoSQL databases is that the NoSQL databases do not support updates and deletes. There are various reasons for this. Many of the applications do not need update and delete operations; rather, different versions of the same data are maintained. For example, in Telecom domains, older call detail records (CDRs) are required for auditing and data mining. In enterprise human resource databases, employee's records are maintained even if an employee may have left the organization. These, updates and deletes are handled using insertion with version control. For example, Bigtable[33] associates a time-stamp with every data item. Further, one needs customized techniques for implementing efficient joins over massive scale data in distributed settings. Thus, joins are avoided in NoSQL databases.

Relational databases provide ACID (Atomicity, consistency, integrity, and durability) properties which may be more than necessary for various data management applications and use cases. Atomicity of over more than one record is not required in most of the applications. Thus, single key atomicity is provided by NoSQL databases. In traditional databases, strong consistency is supported by using conflict resolution at write time (using read and write locks) which leads to scalability problems. As per [26], databases cannot ensure three CAP properties simultaneously: Consistency, Availability, and Partition tolerance (i.e., an individual operation should complete even if individual components are not available). Among consistency and availability, the later is given more importance by various NoSQL databases, e.g., giving service to a customer is more important. Consistency can be ensured using *eventual consistency* [27] where reconciliation happens asynchronously to have eventually consistent database. Similarly, most applications do not need *serialization isolation* level (i.e., to ensure that operations are deemed to be performed one after the other). *Read committed* (i.e., lock on writes till the end of transaction) with single key atomicity is sufficient. Durability is ensured in traditional relational databases as well as NoSQL databases. But traditional databases provide that by using expansive hardware whereas NoSQL databases provide that with cluster of disks with replication and other fault tolerance mechanisms. An alternative of ACID for distributed NoSQL databases is BASE (Basic Availability, Soft-state, and Eventual consistent). By not following the ACID properties strictly, query processing is made faster for massive scale data. As per [26], one does not have any choice between the two (one has to go for BASE) if one needs to scale up for processing massive amount of data. Next we present various data models used for NoSQL along with their commercial implementations.

3.2 NoSQL Data Models

An important reason of popularity of NoSQL databases is their flexible data model. They can support various types of data models and most of these are not strict. In relational databases, data is modeled using relations and one needs to define schema before one starts using the database. But NoSQL databases can support key-value pairs, hierarchical data, geo-spatial data, graph data, etc., using a simple model. Further, in new data management applications, there is a need to frequently keep modifying schema, e.g., a new service may require an additional column or complex changes in data-model. In relational databases, schema modification is time consuming and hard. One needs to lock a whole table for modifying any index structure. We describe three data-models, namely, *key-value stores*, *document stores*, and *column families*.

- **Key-Value Stores:** In a key-value store, read and write operations to a data item are uniquely identified by its key. Thus, no primitive operation spans multiple data items (i.e., it is not easy to support range queries). Amazon's Dynamo [29] is an example of key-value store. In Dynamo, values are opaque to the system and they are used to store objects of size less than 1 MB. Dynamo provides incremental scalability; hence, keys are partitioned dynamically using a hash function to distribute the data over a set of machines or nodes. Each node is aware of keys handled by its peers allowing any node to forward a key's read or write operation to the correct set of nodes. Both read and write operations are performed on a number of nodes to handle data durability, and availability. Updates are propagated to all the replicas asynchronously (eventual consistency). Kai [25] is open source implementation of key-value store.
- **Document Stores:** In document stores, value associated with a key is a document which is not opaque to the database; hence, it can be queried. Unlike relational databases, in a document store, each document can have a different schema. Amazon's SimpleDB[30], MongoDB[32] and Apache's CouchDB [31] are some examples of NoSQL databases using this model. In CouchDB, each document is serialized in JSON (Java Script Object Notation) format, and has a unique document identifier (*docId*). These documents can be accessed using web-browser and queried using JavaScripts. As this database does not support any delete or update; in each read operation multiple versions are read and the most recent one is returned as the result. CouchDB supports real time document transformation and change notifications. The transformations can be done using the user provided *map* and *reduce* JavaScript functions (explained later in the chapter).

- **Column family stores:** Google's BigTable [33] pioneered this data model. BigTable is a sparse, distributed, durable, multi-dimensional sorted map (i.e., sequence of nested key-value pairs). Data is organized into tables. Each record is identified by a row key. Each row has a number of columns. Intersection of each row and column contains time-stamped data. Columns are organized into column-families or related columns. It supports transactions under a single row key. Multiple-row transactions are not supported. Data is partitioned by sorting row-keys lexicographically. BigTable can serve data from disk as well as memory. Data is organized into tablets of size 100-200 MB by default with each tablet characterized by its start-key and end-key. A tablet server manages 10s of tablets. Meta-data tables are maintained to locate tablet-server for a particular key. These metadata tables are also split into tablets. A *chubby file* is the root of this hierarchical meta-data, i.e., this file points to a root metadata tablet. This root tablet points to other metadata tablets which in turn points to user application data tablets. HBase [18] is an open source implementation of BigTable.

Table 1. Comparison of NoSQL databases with traditional relational databases

Product/feature	Dynamo	CouchDB	BigTable	Traditional Databases
Data Model	Key value rows	Documents	Column store	Rows/Relational
Transactional access	Single tuple	Multiple documents	Single and range	Range, complex
Data partition	Random	Random	Ordered by key	Not applicable
Consistency	Eventual	Eventual	Atomic	Transactional
Version control	Versions	Document version	Timestamp	Not applicable
Replication	Quorum for read and write	Incremental replication	File system	Not applicable

Table 1 provides comparison of these different types of NoSQL databases with traditional relational databases. Next we present Hadoop technology which can be used for decision support processing in a warehouse like setting.

4   Hadoop MapReduce

Google introduced MapReduce [6] framework in 2004 for processing massive amount of data over highly distributed cluster of machines. It is a generic framework to write massive scale data applications. This framework involves writing two user defined generic functions: *map* and *reduce*. In the *map* step, a master node takes the input data and the processing problem, divides it into smaller data chunks and sub-problems; and distributes them to worker nodes. A worker node processes one or more chunks using the sub-problem assigned to it. Specifically, each *map* process, takes a set of {*key*, *value*} pairs as input and generates one or more intermediate {*key, value*} pairs for each input key. In the *reduce* step, intermediate key-value pairs are processed to produce the output of the input problem. Each *reduce* instance takes a *key* and an array of *values* as input and produces output after processing the array of *values*:

$$Map(k_1, v_1) \bullet list(k_2, v_2)$$
$$Reduce(k_2, list(v_2)) \bullet list(v_3)$$

Figure 2 shows an example MapReduce implementation for a scenario where one wants to find the list of customers having total transaction value more than $1000.

```
void map(String rowId, String row):
   // rowId: row name
   // row: a transaction recode
  customerId= extract customer-id from row
  transactionValue= extract transaction value from row
  EmitIntermediate(customerId, transactionValue);

  void reduce(String customerId, Iterator partialValues):
    // customerId: Id to identify a customer
    // partialValues: a list of transaction values
    int sum = 0;
    for each pv in partialValues:
      sum += pv;
    if(pv > 1000)
      Emit(cutsomerId, sum);
```

Fig. 2. Example *MapReduce* code

4.1 Hadoop

Hadoop [4] is the most popular open source implementation of MapReduce framework [6]. It is used for writing applications processing vast amount of data in parallel on large clusters of machines in a fault-tolerant manner. Machines can be added and removed from the clusters as and when required. In Hadoop, data is stored on Hadoop Distributed File System (HDFS) which is a massively distributed file system designed to run on cheap commodity hardware. In HDFS, each file is chopped up into a number of blocks with each block, typically, having a size of 64MB. As depicted in Figure 3, these blocks are parsed by user-defined code into {*key, value*} pairs to be read by *map* functions. The *map* functions are executed on distributed machines to generate output {*key, value*} pairs which are written on their respective local disks. The whole *key* space is partitioned and allocated to a number of reducers. Each *reduce* function uses HTTP GET method to pull {*key, value*} pairs corresponding to its allocated key space. For each key, a reduce instance processes the *key* and array of its associated *values* to get the desired output. HDFS follows master-slave architecture. An HDFS cluster, typically, has a single master, also called name node, and a number of slave nodes. The name node manages the file system name space, divides the file into blocks, and replicates them to different machines. Slaves, also called data nodes, manage the storage corresponding to that node. Fault tolerance is achieved by replicating data blocks over a number of nodes. The master node monitors progress of data processing slave nodes and, if a slave node fails or it is slow, reassigns the corresponding data-block processing to another slave node. In Hadoop, applications can be written as a series of MapReduce tasks also. Authors of [9] provide various data models one can use for efficient processing of data using MapReduce, such as universal model [11], column store [10], etc. By avoiding costly joins and disk reads, a combination of universal data and column store proves to be the most efficient data model.

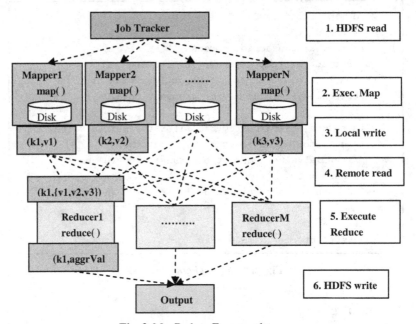

Fig. 3. MapReduce Framework

5   Data Stream Management System (DSMS)

We next turn our attention towards the second significant component of the Big Data story: analysis of the data in motion. In a conventional database, a query executes once and returns a set of results. In contrast in a streaming scenario, a query keeps getting continuously executed over a stream of data. Thus, rather than gathering large quantities of data, storing it on disk, and then analyzing it, data stream management systems (DSMSs) allow user to analyze the data-in-motion. This analysis is done in real-time thereby allowing users to trigger important events to enable enterprises to perform actions *just-in-time* yielding better results for the businesses. There are various enterprise class stream processing systems available. These systems provide better scalability and higher throughput compared to complex event processing (CEP) systems. Distributed stream processing systems process input streams by transforming the tuples in a distributed manner using a system of computing elements and produce output streams. IBM InfoSphere Streams [36], S4 [35] and Storm [34] are some of the examples of such systems. These systems are particularly geared towards clusters of commodity hardware. Thus, one can use cloud infrastructure to perform various stream operations. These systems can be thought of as a series of connected operators. *Source operators* represent sources of data tuples. Intermediate operators perform various operations such as filtering, window aggregation, join, etc. Finally, output is fed to the *sink operators*. We describe these operators in details in the next section.

5.1   Various Stream Processing Systems

We describe three steam processing systems in this section: IBM's InfoSphere Streams, Twitter's Storm, and Yahoo's S4. InfoSphere Streams is a component based distributed stream processing platform, build to support higher data rates and various input data types. It also provides scheduling, load balancing, and high availability to ensure needs for scalability. Streams offers three methods for end-users to operator on streaming data: 1) Stream processing application declarative engine (SPADE) provides a language and runtime to create applications without understanding lower-level operations; 2) User queries can be expressed as per their information needs and interests, which are automatically converted into set of application components; 3) User can develop applications through an integrated development environment (IDE).

Storm provides with a general framework for performing streaming computations, much like Hadoop provides programmers with a general framework for performing batch operations. Operator graph defines how a problem should be processed and how data is entered and processed through the system by means of data streams. *Spouts* are entities that handle the insertion of data tuples into the topology and *bolts* are entities that perform computation. The *spouts* and *bolts* are connected by streams to form a directed graph. Parts of the computation can be parallelized by setting up a parallelism number for each *spout* and *bolt* in the job specification. *Bolts* can also send tuples to external systems, e.g., distributed databases or notification services. Storm supports the major programming languages to encode *spouts* and *bolts*. Storm supports acknowledge based guaranteed communication and generalized stream which takes any kind of objects and primitives (e.g., using thrift). Storm provides fault tolerance just like Hadoop in the face of failures of one or more operators.

In S4 terminology, each stream is described as a sequence of events having pairs of keys and attributes. Basic computational units in S4 are processing elements (PEs). Each instance of a PE is uniquely identified by the functionality of the PE, types of events that the PE consumes, the keyed attributes in those events, and the value of the keyed attributes in the events. Each PE consumes exactly those events which correspond to the value on which it is keyed. Processing nodes (PNs) are logical host for PEs. S4 routes every event to PNs based on a hash function of the values of all known keyed attributes in that event. Its programming paradigm includes writing generic, reusable, and configurable PEs which can be used across various applications. In the event of failures or higher rate events, it degrades performance by eliminating events as explained in the next section.

5.2 Example Stream Processing Algorithms

In a stream processing scenario, we need to generate answers quickly without storing all the tuples. In a typical stream processing application, some summary information of past seen tuples is maintained for processing future data. Since, the data processing is required to be done in (near) real-time, one uses in-memory storage for the summary information. With a bounded amount of memory, more often than not, it is not always possible to produce exact answers for data stream queries. In comparison, data-at-rest almost always produces exact answer (although there are some works giving on-line answers for long running Hadoop jobs). There are various works in the literature providing high quality approximation algorithms (with approximation guarantees) over data streams [38]. Various sampling techniques are proposed for matching streaming data rates with the processing rate. Random sampling can be used as simplest form of summary structure where a small sample is expected to capture the essential features of the stream. Sketching [37] is very popular technique for maintaining limited randomized summary of data for distributed processing of streams. Such sketches have been used for calculating various frequency counts (e.g., estimating number of distinct values in the stream). Histogram is a commonly used structure to capture the data distribution. Histograms can be used for query result size estimation, data mining, etc. Equi-width histograms, end-biased histograms, etc., are various types of histograms proposed in the literature. End-biased histograms can be used to answer Iceberg queries.

6 Querying Data over Cloud

In the previous section we discussed how we can process data using NoSQL databases, Hadoop, and DSMS. Various NoSQL databases are accessed using *get(key)* methods. For processing data in Hadoop one needs to write MapReduce programs. A MapReduce program can be written in various programming languages such as Java, Python, Ruby, etc. But this approach of writing custom MapReduce functions for each application has many problems:

1. Writing custom MapReduce jobs is difficult and time consuming and requires highly skilled developers.
2. For each MapReduce job to run optimally, one needs to configure a number of parameters, such as number of *mappers* and *reducers*, size of data block each *mapper* will process, etc. Finding suitable values of these parameters is not easy.
3. An application may require a series of MapReduce jobs. Hence, one needs to write these jobs and schedule them properly.
4. For efficiency of MapReduce jobs one has to ensure that all the reducers get a similar magnitude of data to process. If certain reducers get a disproportionate magnitude of data to process, these reducers will keep running for a long period of time while other reducers are sitting idle. This will hence in turn impact the performance of the MapReduce program.

Thus, instead various high level query languages have been developed so that one can avoid writing low level MapReduce programs. Queries written in these languages are in turn translated into equivalent MapReduce jobs by the compiler and these jobs are then consequently executed on Hadoop. Three of these languages Hive [7, 20], Jaql [13, 14], and Pig [16, 17] are the most popular languages in Hadoop Community. An analogy here would be to think of writing a MapReduce program as writing a Java program to process data in a relational database; while using one of these high level languages is like writing a script in SQL. We next briefly discuss each of these.

6.1   High Level Query Languages for Hadoop

We next briefly outline the key features of these three high level query languages to process the data stored in HDFS. Table 2 provides a comparison of these three scripting languages [3].

1. Hive: Hive [7,20] provides an easy entry point for data analysts, minimizing the effort required to migrate to the cloud based Hadoop infrastructure for distributed data storage and parallel query processing. Hive has been specially designed to make it possible for analysts with strong SQL skills (but meager Java programming skills) to run queries on huge volumes of data. Hive provides a subset of SQL, with features like *from* clause sub-queries, various types of *joins*, *group-by*s, *aggregation*s, *"create table as select"*, etc. All these features make Hive very SQL-like. The effort required to learn and to get started with Hive is pretty small for a user proficient in SQL.

Hive structures data into well-understood database concepts like tables, columns, rows, and partitions. The schema of the table needs to be provided up-front. Just like in SQL, a user needs to first create a table with a certain schema and then only the data consistent with the schema can be uploaded to this table. A table can be partitioned on a set of attributes. Given a query, Hive compiler may choose to fetch the data only from certain partitions, and hence, partitioning helps in efficiently answering a query. It supports all the major primitive types: *integer*, *float*, *double* and *string*, as well as collection types such as *map*, *list* and *struct*. Hive also includes a system catalogue, a meta-store, that contains schemas and statistics, which are useful in data exploration, query optimization and query compilation [20].

2. Pig: Pig is a high-level scripting language developed by Yahoo to process data on Hadoop and aims at a sweet spot between SQL and MapReduce. Pig combines the best of both-worlds, the declarative style of SQL and low level procedural style of MapReduce. A Pig program is a sequence of steps, similar to a programming language, each of which carries out a single data transformation. However the transformation carried out in each step is fairly high-level e.g., filtering, aggregation etc., similar to as in SQL. Programs written in Pig are firstly parsed for syntactic and instance checking. The output from this parser is a logical plan, arranged in a directed acyclic graph, allowing logical optimizations, such as projection pushdown to be carried out. The plan is compiled by a MapReduce compiler, which is then optimized once more by a MapReduce optimizer performing tasks such as early partial aggregation. The MapReduce program is then submitted to the Hadoop job manager for execution.

Pig has a flexible, fully nested data model and allows complex, non-atomic data types such as set, map, and tuple to occur as fields of a table. A *bytearray* type is supported, to facilitate unknown data types and lazy conversion of types. Unlike Hive, stored schemas are optional. A user can supply schema information on the fly or can choose not to supply at all. The only capability required is to be able to read and parse the data. Pig also has the capability of incorporating user define functions (UDFs). A unique feature of Pig is that it provides a debugging environment. The debugging environment can generate a sample data to help a user in locating any error made in a Pig script.

3. Jaql: Jaql is a functional data query language, designed by IBM and is built upon JavaScript Object Notation (JSON) [8] data model. Jaql is a general purpose data-flow language that manipulates semi-structured information in the form of abstract JSON values. It provides a framework for reading and writing data in custom formats, and provides support for common input/output formats like CSVs, and like Pig and Hive, provides operators such as filtering, transformations, sort, group-bys, aggregation, and join. As the JSON model provides easy migration of data to- and from- some popular scripting languages like JavaScript and Python, Jaql is extendable with operations written in many programming languages. JSON data model supports atomic values like numbers and strings. It also supports two container types: *array* and *record* of name-value pairs, where the values in a container are themselves JSON values. Databases and programming languages suffer an impedance mismatch as both their computational and data models are so different. As JSON has a much lower impedance mismatch (with respect to Java) than, XML for example, but has much richer data types than relational tables. Jaql comes with a rich array of built-in functions for processing unstructured or semi-structured data as well. For example, Jaql provides a bridge for *SystemT* [14] using which a user can convert natural language text into a structured format. Jaql also provides a user with the capability of developing *modules*, a concept similar to Java *packages*. A set of related functions can be bunched together to form a module. A Jaql script can import a module and can use the functions provided by the module.

Table 2. Comparison of Hive, Pig, and Jaql

Feature	Hive	Pig	Jaql
Developed by	Facebook	Yahoo	IBM
Specification	SQL like	Data flow	Data flow
Schema	Fixed schema	Optional schema	Optional schema
Turning complete-ness	Need extension using Java UDF	Need extension using Java UDF	Yes
Data model	Row oriented	Nested	JSON, XML
Diagnostics	Show, describe	Describe, explain commands	Explain
Java connectivity	JDBC	Custom library	Custom library

6.2   Querying Streaming Data

We describe two query languages for processing the streaming data: Continuous Query Language (CQL) [21] described by STREAM (stanford stream data manager) and Stream processing language (SPL) used with IBM InfoSphere Stream[22].

1. Continuous Query Language: CQL is an SQL based declarative language for continuously querying streaming and dynamic data, developed at Stanford University. CQL semantics is based on three classes of operator: *stream-to-relation*, *relation-to-relation*, and *relation-to-stream*. In *stream-to-relation* operators, CQL has three classes of sliding window operators: *time-based*, *tuple-based*, and *partitioned*. In the first two window operators, window size is specified using a time-interval T and the number of tuples N, respectively. The *partitioned* window operator is similar to SQL group-by which groups N tuples using specified attributes as keys. All *relation-to-relation* operators are derived from traditional relational queries. CQL has three relation-to-stream operators: *Istream*, *Dstream* and *Rstream*. Applying an *Istream/Dstream* (insert/delete stream) operator to a relation R results in a stream of tuples inserted/deleted into/from the relation R. The *Rstream* (relation steam) generates a stream element $\langle s, \tau \rangle$ whenever tuple s is in relation R at time τ. Consider the following CQL statement for filtering a stream:

> Select Istream(*)
> from SpeedPosition [Range Unbounded]
> where speed > 55

This query contains three operators: an *Unbounded* windowing operator producing a relation containing all the speed-position measurements up-to current time; relational filter operator restricting the relations with measurements having speed greater than 55; and *Istream* operator streaming new values in the relation as the continuous query result. It should be noted that there are no *stream-to-stream* operators in CQL. One can generate output streams using input streams by combination of other operators as exemplified above.

When a continuous query is specified in CQL stream management system, it is compiled into a query plan. The generated query plan is merged with existing query plans for sharing computations and storage. Each query plan runs continuously with three types of components: *operators*, *queues*, and *synopses*. Each operator reads from input queues, processes the input based on its semantics, and writes output to output queues. Synopses store the intermediate stage needed by continuous query plans. In CQL query plans synopses for an operator are not used by any other operator. For example, to perform window-join across two streams, a join operator maintains one synopsis (e.g., hash of join attributes) for each of the join inputs.

2. Stream Processing Language: This is a structured application development language to build applications over InfoSphere streams. System-S [23] is the stream processing middleware used by SPL. It supports structured as well as unstructured data stream processing. It provides a toolkit of operators using which one can implement any relational query with window extensions. It also supports extensions for application domains like signal processing, data mining, etc. Among operators, *functor* is used for performing tuple level operations such as filtering, projection, attribute creation, etc.; *aggregate* is used for grouping and summarization; *join* is used for correlating two streams; *barrier* is used for consuming tuples from multiple streams and outputting a tuple in a particular order; *punctor* is also for tuple level manipulations where conditions on the current and past tuples are evaluated for generating punctuations in the output stream; *split* is used for routing tuples to multiple output streams; and *delay* operator is used for delaying a stream based on a user-supplied time interval. Besides these System-S also has edge adaptors and user defined operators. A *source* adaptor is used for creating stream from an external source. This adaptor is capable of parsing, tuple creation, and interacting with diverse external devices. A *sink* adaptor can be used to write tuples into a file or a network. It supports three types of windowing: *tumbling* window, *sliding* window, and *punctuation*-based window. Its application toolkit can be quickly used by application developers for quickly prototyping a complex streaming application.

7   Data Management Applications over Cloud

In this section, we consider three example applications where large scale data management over cloud is used. These are specific use-case examples in telecom, finance, and sensors domains. In the telecom domain, massive amount of *call detail records* can be processed to generate near real-time network usage information. In finance domain we describe the fraud detection application. We finally describe a use-case involving massive scale spatio-temporal data processing.

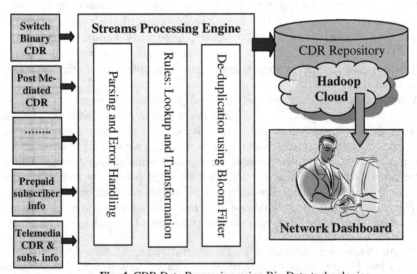

Fig. 4. CDR Data Processing using Big Data technologies

7.1   Dashboard for CDR Processing

Telecom operators are interested in building a dashboard that would allow the analysts and architects to understand the traffic flowing through the network along various dimensions of interest. The traffic is captured using *call detail records* (CDRs) whose volume runs into a terabyte per day. CDR is a structured stream generated by the telecom switches to summarize various aspects of individual services like voice, SMS, MMS, etc. Monitoring of CDRs flowing through cell sites helps the telecom operator decide regions where there is high network congestion. Adding new cell sites is the obvious solution for congestion reduction. However, each new site costs more than 20K USD to setup. Therefore, determining the right spot for setting up the cell site and measuring the potential traffic flowing through the site will allow the operator to measure the return on investment. Other uses of the dashboard include determining the cell site used most for each customer, identifying whether users are mostly making calls within cell site calls, and for cell sites in rural areas identifying the source of traffic i.e. local versus routed calls. Given the huge and ever growing customer base and large call volumes, solutions using traditional warehouse will not be able to keep-up with the rates required for effective operation. The need is to process the CDRs in near real-time, mediate them (i.e., collect CDRs from individual switches, stitch, validate, filter, and normalize them), and create various indices which can be exploited by dashboard among other applications. An IBM SPL based system leads to mediating 6 billion CDRs per day [24]. The dashboard creates various aggregates around combinations of the 22 attributes for helping the analysts. Furthermore, it had to be projected into future based on trends observed in the past. These CDRs can be loaded periodically over cloud data management solution. As cloud provides flexible storage, depending on traffic one can decide on the storage required. These CDRs can be processed using various mechanisms described in the chapter to get the required key performance indicators. Figure 4 shows schematic of such a system.

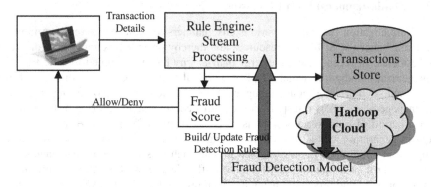

Fig. 5. Credit card fraud detection using Big Data technologies

7.2   Credit Card Fraud Detection

More than one-tenth of world's population is shopping online [2]. Credit card is the most popular mode of online payments. As the number of credit card transactions rise, the opportunities for attackers to steal credit card details and commit fraud are also increasing. As the attacker only needs to know some details about the card (card number, expiration date, etc.), the only way to detect online credit card fraud is to analyze the spending patterns and detect any inconsistency with respect to usual spending patterns. Various credit card companies compile a consumer profile for each and every card holder based on purchases s/he makes using his/her credit card over the years. These companies keep tabs on the geographical locations where the credit card transactions are made—if the area is far from the card holder's area of residence, or if two transactions from the same credit card are made in two very distant areas within a relatively short timeframe, — then the transactions are potentially fraud transactions. Various data mining algorithms are used to detect patterns within the transaction data. Detecting these patterns requires the analysis of large amount of data. For analyzing these credit card transactions one may need to create tuples of transaction for a particular credit card. Using these tuples of the transactions, one can find the distance between geographic locations of two consecutive transactions, amount of these transactions, time difference between transactions, etc. By these parameters, one can find the potential fraudulent transactions. Further data mining, based on a particular user's spending profile can be used to increase the confidence whether the transaction is indeed fraudulent.

As number of credit card transactions is huge and the kind of processing required is not a typical relational processing (hence, warehouses are not optimized to do such processing), one can use Hadoop based solution for this purpose as depicted in Figure 5. Using Hadoop one can create customer profile as well as creating matrices of consecutive transactions to decide whether a particular transaction is a fraud transaction. As one needs to find the fraud with-in some specified time, stream processing can help. By employing massive resources for analyzing potentially fraud transactions one can meet the response time guarantees.

7.3   Spatio-temporal Data Processing

Rapid urbanization has caused various city planners to rethink and redesign how we can make better use of limited resources and thereby provide a better quality of life to the residents. In recent years, the term *smarter city* has been coined to refer to a city management paradigm in which IT plays a crucial role. A smart city promises to bring greater automation, intelligent routing and transportation, better monitoring, and better city management. Smart transportation requires continuous monitoring of the vehicles over a period of time to gain patterns of behavior of traffic and road incidents. This requires generating, collecting, and analyzing massive data which is inherently spatio-temporal in nature. For monitoring and mining such massive amount of dynamic data we need big data technologies. Real time traffic data as well as weather information can be collected and processed using a stream processing system. Any traveler can get the best route for a particular destination by querying using her location and the destination. Another example is smarter environment monitoring system. Such a system is envisioned to collect weather, seismic, and pollution data. Such sensors are deployed all over a city and generate a number of readings everyday. Such data can be used to locate the source of an air pollution incident where air dispersion models can be run to provide information on probable locations of the pollution source.

8   Discussion and Conclusion

We presented the need for processing large amount of data having high variety and veracity at high speed; different technologies for distributed processing such as NoSQL, Hadoop, Streaming data processing; Pig, Jaql, Hive, CQL, SPL for querying such data; and customer use cases. There are various advantages in moving to cloud resources from dedicated resources for data management. But some of the enterprises and governments are still skeptical about moving to cloud. More work is required for cloud security, privacy and isolation areas to alleviate these fears. As noted earlier various applications involve huge amount of data and may require real time processing, one needs tools for bulk processing of huge amount of data, real time processing of streaming data and method of interaction between these two modules. For given cloud resources one needs to associate required resources for both the modules (bulk and stream data processing) so that the whole system can provide the required response time with sufficient accuracy. More research is required for facilitating such systems.

References

1. Avizienis, A.: Basic concepts and taxonomy of dependable and secure computing. IEEE Transactions on Dependable and Secure Computing (2004)
2. Srivastava, A., Kundu, A., Sural, S., Majumdar, A.: Credit Card Fraud Detection using Hidden Markov Model. IEEE Transactions on Dependable and Secure Computing (2008)
3. Stewart, R.J., Trinder, P.W., Loidl, H.-W.: Comparing High Level MapReduce Query Languages. In: Temam, O., Yew, P.-C., Zang, B. (eds.) APPT 2011. LNCS, vol. 6965, pp. 58–72. Springer, Heidelberg (2011)
4. Apache Foundation. Hadoop, http://hadoop.apache.org/core/
5. Awadallah, A.: Hadoop: An Industry Perspective. In: International Workshop on Massive Data Analytics Over Cloud (2010) (keynote talk)
6. Dean, J., Ghemawat, S.: MapReduce: Simplified Data Processing on Large Clusters. Communications of ACM 51(1), 107–113 (2008)
7. Hive- Hadoop wiki, http://wiki.apache.org/hadoop/Hive
8. JSON, http://www.json.org
9. Gupta, R., Gupta, H., Nambiar, U., Mohania, M.: Enabling Active Archival Over Cloud. In: Proceedings of Service Computing Conference, SCC (2012)
10. Stonebraker, M., et al.: C-STORE: A Column-oriented DBMS. In: Proceedings of Very Large Databases, VLDB (2005)
11. Vardi, M.: The Universal-Relation Data Model for Logical Independence. IEEE Software 5(2) (1988)
12. Borthakur, D., Jan, N., Sharma, J., Murthy, R., Liu, H.: Data Warehousing and Analytics Infrastructure at Facebook. In: Proceedings of ACM International Conference on Management of Data, SIGMOD (2010)
13. Jaql Project hosting, http://code.google.com/p/jaql/

14. Beyer, K.S., Ercegovac, V., Gemulla, R., Balmin, A., Eltabakh, M., Kanne, C.-C., Ozcan, F., Shekita, E.J.: Jaql: A Scripting Language for Large Scale Semi-structured Data Analysis. In: Proceedings of Very Large Databases, VLDB (2011)
15. Liveland: Hive vs. Pig, http://www.larsgeorge.com/2009/10/hive-vs-pig.html
16. Pig, http://hadoop.apache.org/pig/
17. Olston, C., Reed, B., Srivastava, U., Kumar, R., Tomkins, A.: Pig-Latin: A Not-So-Foreign Language for Data Processing. In: Proceedings of ACM International Conference on Management of Data, SIGMOD (2008)
18. HBase, http://hbase.apache.org/
19. Curino, C., Jones, E.P.C., Popa, R.A., Malviya, N., Wu, E., Madden, S., Balakrishnan, H., Zeldovich, N.: Realtional Cloud: A Database-as-a-Service for the Cloud. In: Proceedings of Conference on Innovative Data Systems Research, CIDR (2011)
20. Thusoo, A., Sarma, J.S., Jain, N., Shao, Z., Chakka, P., Zhang, N., Anthony, S., Liu, H., Murthy, R.: Hive – A Petabyte Scake Data Warehouse Using Hadoop. In: Proceedings of International Conference on Data Engineering, ICDE (2010)
21. Arasu, A., Babu, S., Widom, J.: The CQL Continuous Query Language: Semantic Foundations and Query Execution. VLDB Journal (2005)
22. Zikopoulos, P., Eaton, C., Deroos, D., Deutsch, T., Lapis, G.: Understanding Big Data: Analytics for Enterprise Class Hadoop and Streaming Data. McGrawHill (2012)
23. Gedik, B., Andrade, H., Wu, K.-L., Yu, P.S., Doo, M.: SPADE: The System S Declaratve Stream Processing Engine. In: Proceedings of ACM International Conference on Management of Data, SIGMOD (2008)
24. Bouillet, E., Kothari, R., Kumar, V., Mignet, L., et al.: Processing 6 billion CDRs/day: from research to production (experience report). In: Proceedings of International Conference on Distributed Event-Based Systems, DEBS (2012)
25. Kai, http://sourceforge.net/apps/mediawiki/kai
26. Fox, A., Gribble, S.D., Chawathe, Y., Brewer, E.A., Gauthier, P.: Cluster-Based Scalable Network Services. In: Proceedings of the Sixteenth ACM Symposium on Operating Systems Principles, SOSP (1997)
27. Wada, H., Fekede, A., Zhao, L., Lee, K., Liu, A.: Data Consistency Properties and the Trade-offs in Commercial Cloud Storages: the Consumers' Perspective. In: Proceedings of Conference on Innovative Data Systems Research, CIDR (2011)
28. Gray, J., Helland, P., O'Neil, P.E., Shasha, D.: The Dangers of Replication and a Solution. In: Proceedings of ACM International Conference on Management of Data (1996)
29. DeCandia, G., Hastorun, D., Jampani, M., Kakulapati, G., Lakshman, A., Pilch, A., Sivasubramanian, S., Vosshall, P., Vogels, W.: Dynamo: Amazon's Highly Available Key-value Store. In: Proceedings of Twenty-First ACM SIGOPS Symposium on Operating Systems Principles, SOSP (2007)
30. Habeeb, M.: A Developer's Guide to Amazon SimpleDB. Pearson Education
31. Lehnardt, J., Anderson, J.C., Slater, N.: CouchDB: The Definitive Guide. O'Reilly (2010)
32. Chodorow, K., Dirolf, M.: MongoDB: The Definitive Guide. O'Reilly Media, USA (2010)
33. Chang, F., Dean, J., Ghemawat, S., Hsieh, W.C., Wallach, D.A., Burrows, M., Chandra, T., Fikes, A., Gruber, R.E.: BigTable: A Distributed Storage System for Structured Data. In: Proceedings of the 7th USENIX Symposium on Operating Systems Design annd Implementation, OSDI (2006)

34. Storm: The Hadoop of Stream processing, `http://fierydata.com/2012/03/29/storm-the-hadoop-of-stream-processing/`
35. Neumeyer, L., Robbins, B., Nair, A., Kesari, A.: S4: Distributed Stream Computing Platform. In: IEEE International Conference on Data Mining Workshops, ICDMW (2010)
36. Biem, A., Bouillet, E., Feng, H., et al.: IBM infosphere streams for scalable, real-time, intelligent transportation services. In: SIGMOD 2010 (2010)
37. Alon, N., Matias, Y., Szegedy, M.: The space complexity of approximating the frequency moments. In: Proceedings of the Annual Symposium on Theory of Computing, STOC (1996)
38. Babcock, B., Babu, S., Datar, M., Motvani, R., Widom, J.: Model and Issues in Data Streams Systems. ACM PODS (2002)

A Model of Virtual Crop Labs as a Cloud Computing Application for Enhancing Practical Agricultural Education

Polepalli Krishna Reddy[1], Basi Bhaskar Reddy[1], and D. Rama Rao[2]

[1] IT for Agriculture and Rural Development Research Center,
International Institute of Information Technology Hyderabad (IIITH),
Gachibowli, Hyderabad, India
pkreddy@iiit.ac.in
[2] National Academy of Agricultural Research Management (NAARM),
Rajendranagar, Hyderabad, India
ramarao@naarm.ernet.in

Abstract. A model of crop specific virtual labs is proposed to improve practical agricultural education by considering the agricultural education system in India. In agricultural education, the theoretical concepts are being imparted through class room lectures and laboratory skills are imparted in the dedicated laboratories. Further, practical agricultural education is being imparted by exposing the students to the field problems through Rural Agricultural Work Experience Program (RAWEP), experiential learning and internships. In spite of these efforts, there is a feeling that the level of practical skills exposed to the students is not up to the desired level. So we have to devise the new ways and means to enhance the practical knowledge and skills of agricultural students to understand the real-time crop problems and provide the corrective steps at the field level. Recent developments in ICTs, thus, provide an opportunity to improve practical education by developing virtual crop labs. The virtual crop labs contain a well organized, indexed and summarized digital data (text, photograph, and video). The digital data corresponds to farm situations reflecting life cycles of several farms of different crops cultivated under diverse farming conditions. The practical knowledge of the students could be improved, if we systematically expose them to virtual crop labs along with course teaching. We can employ cloud computing platform to store huge amounts of data and render to students and other stakeholders in an online manner.

Keywords: IT for agriculture education, agro-informatics, virtual crop labs, decision support system.

1   Introduction

The crop problems are highly location/resource specific and vary considerably with agro-climatic zones. The crop production and protection related problems

S. Srinivasa and V. Bhatnagar (Eds.): BDA 2012, LNCS 7678, pp. 62–76, 2012.

vary based on variations in soil, seed, weather, cropping history, crop environment and agro-climatic zone. Also, the problems vary year to year due to seasonal variations.

It is expected that the graduates produced from agricultural universities should possess theoretical, lab and practical (field) knowledge regarding crop husbandry. The theoretical knowledge includes fundamental concepts regarding crop husbandry. Laboratory knowledge cover the ability to diagnose the problem in the dedicated lab once the samples are provided. The practical knowledge includes the ability to (i) understand the seed to seed spatio-temporal problem dynamics of farms in diverse agro-climatic situations for different crops (crop varieties) and (ii) give advisories/suitable corrective measures for crop various husbandry problems.

We have made an effort to propose a model of virtual crop labs to impart practical agricultural education by considering the agricultural education system in India. However, the model is general enough to be extended to other countries having agricultural farming and education systems similar to India.

In agricultural education, the theoretical concepts are being imparted through class room lectures and the laboratory skills in the dedicated laboratories. Regarding imparting of practical knowledge, efforts are being made to give practical education by exposing the students to the field problems in college farms and the program called Rural Agricultural Work Experience Programme (RAWEP). In spite of these efforts, still there is a feeling that the level of practical skills exposed to the students is not up to the desired level. So, we have to devise the methods to enhance the practical knowledge and skills of agricultural students to understand the real time crop problems and provide the corrective steps at the field level. In this paper, we propose a framework of virtual crop labs for enhanced practical education by exploiting the developments in information and communication technologies (ICTs) such as database, digital photography, video and internet technology.

In the next section, we explain the related work. In section 3, we discuss the importance of practical education and gaps in the existing practical learning framework. In section 4, we present the proposed model of virtual crop labs. In section 5, we discuss the role of virtual crop labs in practical learning. In section 6, we explain how the proposed framework results into huge data maintenance and rendering task and requires cloud computing environment. The last section contains conclusions and future work.

2   Related Work

DIAGNOSIS in plant pathology - teaching tool that has been used at Massey University and other institutions in New Zealand since 1989. DIAGNOSIS presents students with a problem to solve, captures their diagnosis, justification and solution skills, and provides feedback. Students generally react very positively to the exercises. They are forced to integrate and synthesize material learned elsewhere in other classes (soil science, plant physiology, entomology etc.) thereby look and understand the problem in a holistic manner [3]. The real value of DIAGNOSIS is in the quality of the scenarios and how the exercises are used within a course. This quality and appropriateness of use comes from dedicated plant pathology teachers and practitioners themselves. The experience, wisdom and knowledge of extension plant pathologists can be embodied in a scenario, and used by a skilful teacher and pose challenge to students [6][7][8] for identification.

It has been noted in many locations around the globe that applied crop production and protection specialists are not being trained in the numbers needed to support current and future extension efforts in agriculture. Therefore, plant pathology graduate students need to be exposed to applied aspects of plant pathology and especially diagnostic reasoning [9]. Techniques of isolation and pathogen identification are easy to teach in a standard university course, but teaching the skills of the field diagnosis are a lot more difficult, especially in departments without active extension programme and with stretched teaching resources [5]. The same is the case with other applied disciplines of crop husbandry.

A distance diagnostic and identification system (DDIS) was developed at the University of Florida which allows the users to submit digital samples obtained in the field for rapid diagnosis and identification of plants, diseases, insects and animals [10]. Along with DDIS a user friendly rich internet digital media library application called DDISMedia is also being developed to help users to collect, manage, and retrieve high-quality media from the DDIS database and to assist specialists in pest diagnosis. The DDISMedia is a peer-reviewed media database, which contains a collection of digital media of plant pests, diseases and plant samples produced by specialists of different disciplines. The media library could also be used for research, educational programmes, teaching and learning [11].

Fourth's Deans committee constituted by Indian Council of Agricultural Research (ICAR) in 2007 [1] recommended the curriculum which is being followed in Indian agriculture universities. The curriculum stresses the importance of imparting practical education in agriculture. In the research report [2] submitted by National Research Council, USA, the importance of developing innovative methods for imparting practical education in agriculture is stressed.

A system called eSagu [4] has been built in which the agricultural experts have to identify the production/protection/other problems by looking at the digital photographs. It was observed that if expert is provided with dynamics of case and effective relationships of the problems with time, diagnosis as well as advisory delivery becomes easy. Based on this observation, an effort [12] has been made to develop a *data warehouse of crop care concepts* to impart practical learning. Effort has been made to develop framework of virtual crop labs in [13]. In [14], a content development framework has been proposed which contains the content related to location-specific weather sensitive crop operations to improve the quality of agromet bulletins.

3 About Practical Education in Agriculture

In this section, we explain the importance of practical education and limitations of existing practical education framework.

3.1 Importance of Practical Education

To identify the crop problems, the agricultural scientists (plant doctors) are supposed to move themselves to the cropped field, observe the crop critically, raise the questions themselves and diagnose the problem /ailment by following elimination process of cause and effect relationships in crop husbandry. Finally, plant doctors confirm the diagnosed problem if necessary, by using some of the laboratory tests. Thus, the agriculture graduates should posses expertise and confidence to give the agro-advice to the farmer for curing the aliment in the cropped field. The advice may be implemented fully/ partially (faulty) and the reaction of the crop may be different in different farming situations making the correct decision at field level is much more difficult for fresh graduates who have least practical field exposure. Hence, agri-professionals require in depth understanding of the field problems (crop production / protection /post harvest handling / processing) besides intensive exposure of realtime field problems in diversified farming situations. In addition, the demand from the stakeholders (such as agro-based industries) is fast changing with a change in technologies developed in different agro-climatic zones of the country.

The agriculture curriculum which is being followed by all agricultural and horticultural and other related universities in India has been recommended by the Fourth's Deans committee constituted by ICAR, Government of India in 2007 [1]. The report contains norms, standards, academic regulations, UG curricula and syllabi. In the report, the objectives of revised curriculum for BSc (agriculture) are given as follows (Page no.16 of the report).

- To train manpower with more practical orientation to cater to the needs of public, private and corporate sectors in agriculture.
- To impart knowledge and training in interdisciplinary production oriented courses in agriculture.
- To provide experiential learning and hands - on training for developing entrepreneurial skill for being job provider rather than job seekers.

Overall, the report has stressed the importance on practical orientation. Based on the recommendations, the ongoing learning framework is conceptualized. The main features include inclusion of new areas and increase in practical content through experiential learning duration of which is ranging from six months to a year for BSc in agriculture program. In all disciplines, attachment with industry has been indicated.

The learning process in agriculture related courses under the existing framework is depicted in Figure 1. After entering four year under-graduate program, the students gain basic knowledge and skills through courses and labs (refer (A) in Figure 1), and the practical knowledge through RAWEP, college farms, experiential learning and internships (refer (B) in Fig.1). In this framework, enhanced learning happens due to mutual re-enforcement between A and B. During BSc(agriculture) program, the students spend first three years for learning basic knowledge in the classrooms and dedicated labs. As a part of special course or course assignment, the students visit the selected college farms carry out the assignments and gain practical knowledge. To expose the student to real field problems, he/she has to complete RAWEP during the first semester of the fourth year. The students are expected to stay in the village and attached to the host farmer. They should record all the farm operations on the crops cultivated by the host farmer. As a result, they gain the practical knowledge (which are applied aspects of theoretical knowledge) for decision making about the intricacies of various farm operations, social/cultural aspects of crop cultivation, production/ protection problems, and ameliorative measures to overcome them. Practical knowledge is also gained during experiential learning and internship programs. After the programs, they return to the university and complete the remaining course work during the eighth semester. During this period, they get more insights by understanding the practical problems through theoretical analysis.

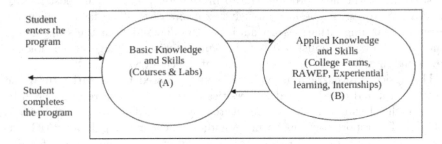

Fig. 1. Depiction of ongoing learning process by the student

3.2   Limitations of Existing Practical Learning Framework

In spite of the best efforts made by agricultural universities and others through changing their curricula /refresher training modules, there is a general feeling that, the level of practical skills exposed to the students / agri-professionals is not up to the desired level. Further, within the limited period of the graduate course programme, it is not feasible to expose the students to a multitude of the crop problems which vary both spatially and temporally. Further, it takes considerable time and effort for an agricultural professional to go to a village, visit the farms and understand the type of practices and problems the farmer faces.

There are also limitations to the RAWEP program. In reality, the students can not afford to spend the sufficient time with the farmer to observe the complete agriculture season as a part of academic program. As a result, they are not able to get exposed to all the crop production, social and cultural practices followed by the farmer from seeding to harvesting of the crop. In addition, the students are not able to stay in the village due to the backlog of courses to be completed and other extra curricular activities such as National Service Scheme (NSS) program of the university. Further, in agriculture, farmers cultivate number of crops in varied micro agro-climatic conditions of a given country. But the host farmer/s in particular village may cultivate 2-3 crops in a given agro-climatic zones and students can get exposure to those crops only either fully or partially. This results in acquisition of limited practical skills/knowledge.

Critical examination of farm problems does reveal that they are highly variable based on space and time. Within a limited period of graduate course programme, it is not feasible to expose the students to a multitude of crop problems. Moreover, the skill set demanded by the agri-industry is also fast changing. Keeping these problems in view, number of committees had been constituted by ICAR and State Agricultural Universities to come up with suitable modifications to the course curricula and internship with industries etc. The major outcome of such exercise is Fourth Dean's Committee report [1] which is being implemented in all state agricultural universities in the country. However, there is a fundamental problem in agriculture due to which nothing much could be done under traditional framework in the agricultural education in the country. Thus, the recent developments in ICT would give new opportunities to impart practical knowledge and revolutionize agricultural education in the country nay in the world.

4   Virtual Crop Labs

Major problem in imparting practical training is the lack of avenues to give adequate exposure of the students to real time crop husbandry practices including farm machinery and power, field preparation, suitable crops, sowing, cropping/farming systems, efficient use of production inputs (soil, water, fertilizer, pesticides), pests, diseases, weeds, harvest, post harvest technologies, special operations (pruning, grafting, nursery raising, seed production including hybrids, poly house technology etc), marketing of agricultural produce and the dynamics of aforesaid activities with space and time. However, by exploiting developments in ICTs, it is possible to capture and organize the crop situations in a virtual manner and expose them to the students as a part of existing curriculum. As a result, there is a scope to improve practical learning.

The main issues in learning practical knowledge are as follows. It can be observed that there are several hundred crops and each crop has several varieties. As a subject matter specialist, an agricultural expert should have practical knowledge on at least few crops. We define the word *practical knowledge* as an ability to understand the plant (crop) health situation or plant (crop) environment (land, water, climate) and come-up with the measures for better growth. On the national-level, there are several macro agro-climatic zones. In addition, each state is divided into several state-level micro agro-climatic zones. For a given crop, the problems and issues are different for different macro and micro agro-climatic zones. If we want to impart the student about the practical knowledge of several crops in different agro-climatic situations, ideally, the student should observe the crop growth for few cycles in these situations and understand a crop husbandry dynamics. So, during four years of graduate program, it is impossible to learn such multitude of crop production problem dynamics in agriculture.

The basic idea of the proposed approach is as follows. We are proposing that it is possible to improve the practical knowledge of students, if we systematically expose them to well organized, indexed and summarized digital data (text, photos and video) of diverse farm situations. So, instead of student going to the crop, the crop growth situation under different agro-climatic situations is captured from seed-to-seed throughout crop growing season at regular intervals covering all the crop husbandry activities. The captured situations are labeled and described by subject-matter specialists (agriculture scientists). The labeled content is exposed to the students of under-graduate programs as a part of learning framework. As a result, a student gets an opportunity to examine the thousands of crop production problem dynamics of each crop and learn the practical knowledge.

To capture diverse farm situations of a crop, different agro-climatic zones will be identified. For each agro-climatic zone, the sample farms are identified by considering soil type and seed variety, and so on. Enough number of sample farms will be selected so that it reflects the diverse farming processes of that zone. For each farm, the crop status and related activities are captured (text, digital photographs and video) at the regular intervals from pre-sowing to post-harvesting. The corresponding problem description and index terms are written by subject matter specialists. In addition, the subject matter specialists also write possible questions and answers for the captured data. So, each virtual crop lab contains crop status of sample farms of several agro-climatic zones in the country.

We propose that we have to develop virtual crop labs. The content of virtual crop labs is systematically exposed to the students during the learning period. The virtual crop lab constitutes *virtual crop lab of zones*. The *virtual crop lab of the zone* constitutes several *virtual farm sagas*. Finally, the *virtual farm saga* constitutes *virtual farm items*. We explain these terms starting from *virtual farm item*.

(i) **Virtual Farm Item (VFI):** Suppose a particular crop (c) is cultivated in the farm (f). We capture the crop situation or activity carried out in the farm at particular instant (or stage) through virtual farm item. So, a *VFI* indicates the farm status of the given farm or activity captured through digital technology. We capture the following elements as a part of VFI: < f, c, d, t, td, p, v, s, q/a, i >. The description of VFI is as follows.

   − *f:* indicates the details of farm (location, ago-climatic situation, soil details, farm size and so on).
   − *c:* indicates the details of crop cultivated in the farm *f* (variety, duration and so on).
   − *d:* indicates the date of sowing of crop *c*.

- *t:* indicates the time (day) of VFI. That is, the farm is visited at time (t) and the values for the parameters *td*, *p* and *v* regarding crop status are captured.
- *td:* indicates description of crop status through text.
- *p:* indicates the set of photographs through which the crop status is captured.
- *v:* indicates the set of video clips through which the crop status is captured.
- *s:* indicates the summary text written by subject matter specialists for the given VFI. It contains the detailed description of the corresponding problem or activity by referring individual photographs or videos. By reading the description, the student or professor can understand the problem or activity captured through VFI.
- *q/a:* indicates questions & answers. The subject matter specialist prepares the questions related to VFI based on the field problems at *t* and provides the answers. These questions are prepared to help the student to get more insights about field problems.
- *i:* indicates the index terms. The index terms will help the information retrieval engine to retrieve the corresponding VFI.

The VFI captures enough information so that the agriculture student/ professor can understand the farm situation or farm activity by going through it in a virtual manner (without visiting the crop field). The situation can be understood through text photograhs/video clips. Based on the crop details and farm location, the student can grasp the problem. To enhance the understanding of crop situation, the student can go through the text written by subject matter specialists. The questions (and answers) should be such that it should enable the student/teacher to understand the background and other influencing factors for that farm situation. Questions are aimed to understand what, how and why aspects of situation/activity, such as the reasons for the given farm situation or importance of such activity.

(ii) **Virtual Farm Saga (VFS):** The VFS captures the activities of a single farm of a given farming situation throughout crop life cycle (from sowing to post-harvest handling) of the crop. For the given farm, the VFS is the collection of virtual farm items which are being captured at the regular intervals from pre-sowing to post-harvesting covering all crop-husbandry activities. The summary and question/anwsers are prepared by considering the VFIs captured during the life cycle of the farm. The structure of VFS is as follows: <f, c, d, Set of VFIs, Summary of VFS, q/a, i>. The description of VFS is as follows.

- *f, c, d, i:* the meaning of these notations is similar as in VFI.
- *Set of VFIs:* Collection VFIs captured at regular intervals from sowing to post-harvesting by covering the crop cycle.
- *Summary of VFS:* The summary is written by considering overall farm situation by considering the corresponding VFIs throughout crop life cycle, i.e., from pre-sowing to post-harvesting. Reasons for the problems

should be mentioned. Best agriculture practices should be highlighted. The negative aspects of crop production practices carried out should be explained. Mistakes identified should be described. Missed opportunities for a better crop growth, if any, should be given.

- q/a: The subject matter specialist should form appropriate questions and give answers regarding dynamics of VFS. Questions can be about the factors about the crop growth, linking activities the problems occurred to the crop to the activities carried out at different stages, weather situation, soil, farm practices, etc.

The VFS captures enough information so that the agriculture student/ professor can understand the growth of the given farm in a virtual manner without visiting the crop field. The student can go through each VFI and understand the crop growth. The questions (and answers) enable the student/teacher to understand the background and other influencing factors for that crop growth. Questions are aimed to understand what influenced the crop growth and corresponding corrective factors that would have been taken, if any.

(iii) **Virtual Crop Lab of a Zone (VCLZ):** The VCLZ captures the crop dynamics for a given micro agro-climatic zone. Let crop c is cultivated in n number of sample farms (in different farming situations) in a given micro agro-climatic zone. The value of n is chosen such that all farming situations of the zone for a given crop are captured. The collection of VFS of n farms (in different farming situations) constitute VCLZ. The elements of VCLZ are as follows: $<c$, Set of VFSs, Summary of VCLZ, q/a, i $>$.

- c, i: the meaning of these notations is similar as in VFI.
- *Set of VFSs:* Indicates a collection of VFSs on n number of sample farms of crop c. For a given agro-climatic zone, the number of sample farms will be selected to capture all the variability (soil, variety, water source and so on) of the crop c.
- *Summary of VCLZ:* The summary for VCLZ is written by considering overall situation of farms in VCLZ. The summary includes reasons for the success of some farms including best agriculture practices, and failure of some other farms which should be mentioned linking to the agro-climatic conditions, good/bad farm practices, crop protection measures, untimely application of inputs, soil type and so on.
- q/a: The questions and answers are formed by subject matter specialists by considering the VFSs of all sample farms. So, there is a scope for several innovative questions and answers which can be given by providing the references to corresponding VFSs and VFIs. Questions can be framed following the processes of contrasting VFSs and VFIs, comparing VFSs and VFIs by identifying unique/special VFSs and VFIs with reference to soil, variety, weather and so on. Questions can also be about the reasons for success of some farms and failure of some other farms linking to seed variety, soil, date of sowing, weather factors and farming practices.

The VCLZ captures enough information so that the agriculture student/ professor can understand the growth of crop dynamics in a virtual manner in the zone under different farming situations. The students compare VFS of several farms and understand the differences in the crop performance under various soils and practices. By going through the summary of VCLZ, the student gets the new insights about the influence of various factors on different types of farms of the same crop in the given zone. The questions (and answers) helps the student to explore the issues in different dimensions.

(iv) **Virtual Crop Lab (VCL):** The virtual crop lab captures all the problems and activities of the crop for all agro-climatic zones of the country. It can be observed that the crop problems differ if the agro-climatic zone changes. So, we develop VCLZ for all different agro-climatic zones of the crop grown in the country. So, the collection of VCLZs of different agro-climatic zones of a crop of the country constitutes VCL. So, the VCL, in effect, captures the diversity of crop problems and activities of several agro-climatic zones in the country.

For each micro agro-climatic zone, we build VCLZ. Let crop c is cultivated in m micro agro-climatic zones in the country. The collection of VCLZs of m zones constitute VCL. The elements of VCL is as follows: $< c,$ Set of VCLZ, Summary of VCL, $q/a, i >$.

- c, i: the meaning of these notations is similar as in VFI.
- *Set of VCLZs:* Indicates VCLZs on m agro-climatic zones.
- *Summary of VCL:* The summary of VCL of a country is written by comparing and contrasting crop growth/farm practices based on the VCLZ of several zones of a country. Reasons for the difference in crop practices by linking to weather and soil should be highlighted.
- *q/a:* The questions and answers are formed by subject matter specialists by considering the VCLZs of several zones. So, there is a scope for several innovative questions and answers can be given by providing the references to corresponding VCLZs, VFSs and VFIs. Questions can be formed following the processes of contrasting VCLZs, VFSs and VFIs, comparing VCLZs, VFSs and VFIs by identifying unique/special VCLZs, VFSs and VFIs with reference to soil, variety, weather and so on. Questions can also be framed about the reasons for the differences in farming practices, crop problems, and protection measures.

The VCL captures enough information so that the agriculture student can understand the growth of crop dynamics in different agro-climatic zones in the country without visiting the fields (virtual manner). The students understand the differences in the crop performance under different agro-climatic zones. By going through the summary of VCL, the student could get the new insights about how crop growth is carried out in different agro-climatic zones, differences in the best agriculture practices in each zone, farming situations in each zone, varietal differences, differences in agronomic practices, dominance of pest and deceases and so on. The questions (and answers) helps the student to explore the crop growth issues in different agro-climatic farming environments.

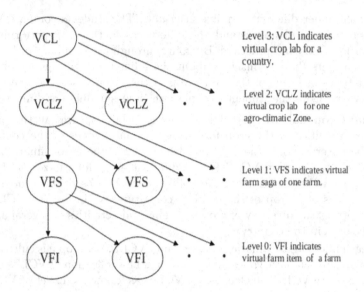

Level 3: VCL indicates
virtual crop lab for a
country.

Level 2: VCLZ indicates
virtual crop lab for one
agro-climatic Zone.

Level 1: VFS indicates virtual
farm saga of one farm.

Level 0: VFI indicates
virtual farm item of a farm

Fig. 2. Relationship among VFI, VFS, VCLZ, VCL. Several VFIs constitute one VFS. Several VFSs constitute one VCLZ. Several VCLZs constitute one VCL.

We have to develop VCL for each crop. It can be observed that the structures VCL, VCLZ, VCS and VFI form an hierarchical relationship (Figure 2). Each *VCL* at *Level 3* is a combination of several VCLZs at *Level 2* and each VCLZ at *Level 2* is a combination of several VCSs at *Level 1*. Finally, each VFS at *Level 1* is a combination of several *VFI* at *Level 0*. Both summaries and q/a should be developed by subject matter specialists at different levels. The nature of summaries and q/a at different levels are given in Table 1.

5   Role of Virtual Crop Labs in Practical Learning

We argue that the virtual crop labs (VCLs) help students and other stakeholders in imparting knowledge about practical learning process. In the ongoing learning framework, the student enters the educational program captures theoretical knowledge during lecture hours and captures practical knowledge by conducting field practicals/experiments on college farms, during RAWEP, experiential learning and internships. So, there is a mutual re-enforcement between theory and practice. In the proposed learning framework, in addition to class room lectures and college farms/RAWEP, the students are exposed to VCLs of crops. Note that VCLs are available to students 24/7 (assuming a laptop/smartphone and internet connectivity is available). Also, for each crop, the VCLs contain several farm situations and activities of several agro-climatic zones. Normally, VCLs also contain farm situations of several years. So, through VCLs, the students and other stakeholders are exposed to crop cultivation and growth in different agro-climatic zones and farming situations. The VCLs provide the opportunity to students and other stakeholders for learning several practical concepts. The teachers can also give several assignments based on the summaries and q/a that were developed as a part of VCLs. The students can gain more practical exposure by completing the assignments through VCLs.

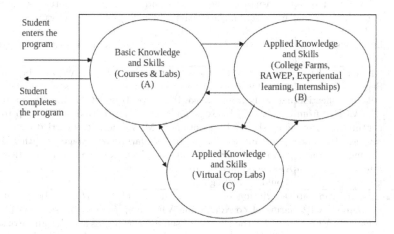

Fig. 3. Depiction of proposed learning process. The mutual re-enforcements A to B and B to A are similar to the existing learning framework. With virtual crop labs, enhanced learning is possible due to the following two kinds of additional mutual re-enforcements: (i) A to C and C to A (ii) B to C and C to B.

Overall, with the addition of virtual crop labs as a part of learning process, the students will gain practical agricultural knowledge through three processes. (A) Classes and labs (B) College farms and RAWEP and (C) Virtual Crop Labs (Figure 3).

(i) **Mutual reinforcement between A and B and vice versa:** This pro-cess is already happening in the ongoing learning process (refer section 3.1 and Fig. 1).

(ii) **Mutual reinforcement between A and C and vice versa:** This pro-cess is an addition to the existing learning system due to virtual crop labs. The knowledge gained during classes/labs can be used to get additional in-sights by studying the farm situations in virtual crop. Also, the practical exposure gained by observing several virtual farm situations can be used to get more theocratical insights.

(iii) **Mutual reinforcement between B to C and vice versa:** This pro-cess is also an addition to the existing learning system due to the proposed virtual crop labs. The farm situations observed in VCLs can be understood by extending the knowledge gained by observing college farms and RAWEP. Also, the knowledge gained during visit to college farms/labs will be enhanced by observing the farm situations in VCLs.

Table 1. Nature of summary and q/a at different levels

Level	Content Type	Nature of Summary	Nature of Questions and Answers (q/a)
Level 0	Virtual Farm Item (VFI) is collection of photographs, videos and text describing farm situation or activity	Description of farm situation or activity such that it is possible to visualize and understand the farm situation/activitivity by going through digital data (text, photos, video).	The q/a are aimed to understand what, how and why aspects of situation/activity such as importance of such activity, reasons for the given farm situation at that stage.
Level 1	Virtual Farm Saga (VFS) constitute collection of VFIs from captured from sowing to harvesting at regular intervals for a given farm.	Description of overall farm situation by considering corresponding VFIs from pre-sowing to post-harvesting. Reasons for the problems should be mentioned. Missed opportunities for a better farm growth, if any, should be given. Also, good farming practices, if any, should be highlighted.	The q/a can be about the factors about the crop growth, linking the problems occurred to the crop to the activities carried out at different stages, weather situation, soil, farm practices.
Level 2	Virtual Crop Lab of a Zone (VCLZ) constitute VFS of sample number of farms for a given zone.	Description of overall situation of farms of corresponding VFS. Reasons for the success of some farms, and failure of some farms should me mentioned linking to the good/bad farm practices, crop protection measures, untimely application, soil type and so on.	The q/a can be about the reasons for success of some farms and the failure of some other farms linking to seed variety, soil, date of sowing, weather factors and farming practices.
Level 3	Virtual Crop Lab (VCL) constitute VCLZs of several zones for a given crop.	The summary is written by comparing and contrasting crop growth/farm practices based on the VCLZ of several zones. Reasons for the difference in crop practices by linking to weather and soil should be highlighted.	The q/a can be about the reasons for the difference in farming practices, crop problems, and protection measures in differet agro-climatic zones.

6   Virtual Crop Labs as a Cloud Computing Application

The virtual crop labs contain the content of hundreds of crops cultivated in the country. For each crop, we capture the crop husbandry dynamics by considering the cultivation/growth under different agro-climatic zones, farming situations, soil types with varying fertility, various levels of water availability, weather types and so on.

The information is useful to all stakeholders such as agriculture students, agriculture extension workers, teachers, farmers, policy planners/makers, researchers and industrialists. The information must be made available in different languages in the country by considering the profiles of different stakeholders in the country.

Developing virtual crop labs and providing access of such information to stakeholders is a major data collection, data storage and data dissemination task. Virtual crop labs contain voluminous data which requires huge disk space for data storage. It also requires a huge bandwidth for streaming the data to large number stakeholders in an online manner in parallel. Such a huge data storage and streaming requirement necessitates the need of porting such application on cloud computing platform.

7   Conclusions and Future Work

In agriculture, it is difficult for agri-professionals to make apt decisions at field level as the practical field problems vary with agro-climatic variability (location), resource variability (crop variety, soil, water, nutrients, energy, management and so on) and temporal variability (season to season and/or year to year due to weather aberrations). With the latest developments in ICTs, it is possible to capture the crop activities and problems both spatially and temporally using the digital technologies and label them. Thus, there is an opportunity to enhance practical learning by exposing such content to agri-professionals in a systematic manner. In this paper we have proposed a framework of virtual crop labs to enhance practical skills. Through virtual crop labs, we capture the crop husbandry dynamics of several crops in a digital form (text, photographs and video). The agriculture students, teachers, agriculture extension workers and other stakeholders can access the information to enhance the practical knowledge. Such a huge data storage and streaming task requires a cloud computing platform.

The proposed model of virtual crop labs is general enough to be extended for any other country. For a country like India, it is a huge task which requires the participation of agriculture/horticulture universities, and ICAR institutes which have been spread over different agro-climatic zones of India. It is a potential application of a cloud computing system. As a part of future work, we are planning to refine the proposed framework and make an effort to build the prototype system to investigate the developmental issues, dissection issues and the corresponding impact on the practical learning aspects of agricultural education.

References

1. Indian Council of Agricultural Research (ICAR): Fourth deans committee on agricultural education in India: report on norms, standards, academic regulations and UG curricula and syllabi, New Delhi (2007)
2. National Research Council Report, Transforming Agriculrural Education for a Changing World (2009)
3. Grogan, R.G.: The science and art of pest and disease diagnosis. Annual Review of Phytopathology 19, 333–351 (1981)
4. Ratnam, B.V., Krishna Reddy, P., Reddy, G.S.: eSagu: An IT based personalized agricultural extension system prototype–analysis of 51 Farmers' case studies. International Journal of Education and Development using ICT (IJEDICT) 2(1) (2006)
5. Schumann, G.L.: Innovations in teaching plant pathology. Annual Review of Phytopathology 41, 377–398 (2003)

6. Stewart, T.M.: Diagnosis. A Microcomputer-based teaching-aid. Plant Disease 76(6), 644–647 (1992)
7. Stewart, T.M., Blackshaw, B.P., Duncan, S., Dale, M.L., Zalucki, M.P., Norton, G.A.: Diagnosis: a novel, multimedia, computer-based approach to training crop protection practitioners. Crop Protection 14(3), 241–246 (1995)
8. Stewart, T.M.: Computer games and other tricks to train plant pathologists. In: Bridge, P.D., Jeffries, P., Morse, D.R. (eds.) Information Technology, Plant Pathology and Biodeversity. CAB International, UK (1997)
9. Stewart, T., Galea, V.: Approaches to training practioners in the art and science of plant disease diagnosis. Plant Disease 90, 539–547 (2006)
10. Xin, J., Beck, H.W., Halsey, L.A., Fletcher, J.H., Zazueta, F.S., Momol, T.: Development of a distance diagnostic and identification system for plant, insect and disease problems. Applied Entineering in Agriculture 17(4), 561–565 (2001)
11. Zhang, S., Xin, J., Momol, T., Zazueta, F.: DDISMedia: A digital media library for pest diagnosis. In: IAALD AFITA WEEA 2008, World Conference on Agricultural Information and IT (2008)
12. Krishna Reddy, P., Shyamasundar Reddy, G., Bhaskar Reddy, B.: Extending ICTs to impart applied agricultural knowledge. In: National Conference on Agro-Informatics and Precision Farming, December 2-3. University of Agricultural Sciences (UAS), Raichur, India (2009)
13. Bhaskar Reddy, B., Krishna Reddy, P., Kumaraswamy, M.: A framework of crop specific virtual labs to impart applied agricultural knowledge. In: Proceedings of Third National Conference on Agro-Informatics and Precision Agriculture 2012 (AIAP 2012), Hyderabad, India, Organized by INSAIT, pp. 69–74. Allied Publishers (2012)
14. Mahadevaiah, M., Raji Reddy, D., Sashikala, G., Sreenivas, G., Krishna Reddy, P., Bhaskar Redddy, B., Nagarani, K., Rathore, L.S., Singh, K.K., Chattopadhyay, N.: A framework to develop content for improving agromet advisories. In: The 8th Asian Federation for Information Technology in Agriculture (AFITA 2012), Taipei, September 3-6 (2012)

Exploiting Schema and Documentation for Summarizing Relational Databases

Ammar Yasir, Mittapally Kumara Swamy, and Polepalli Krishna Reddy

Center for Data Engineering
International Institute of Information Technology-Hyderabad
Hyderabad - 500032, India
yasir@students.iiit.ac.in, kumaraswamy@research.iiit.ac.in,
pkreddy@iiit.ac.in

Abstract. Schema summarization approaches are used for carrying out schema matching and developing user interfaces. Generating schema summary for any given database is a challenge which involves identifying semantically correlated elements in a database schema. Research efforts are being made to propose schema summarization approaches by exploiting database schema and data stored in the database. In this paper, we have made an effort to propose an efficient schema summarization approach by exploiting database schema and the database documentation. We propose a notion of table similarity by exploiting referential relationship between tables and the similarity of passages describing the corresponding tables in the database documentation. Using the notion of table similarity, we propose a clustering based approach for schema summary generation. Experimental results on a benchmark database show the effectiveness of the proposed approach.

Keywords: Schema, Schema Summarization, Database Usability.

1 Introduction

Generating schema summary for a complex relational database schema is a research issue. According to a recent study, the time taken by users to express their query requirements is O(min), while the time taken for executing the query and result display is O(sec) [1]. With the increase in complexity of modern day databases, users spend considerable amount of time in understanding a given schema, in order to locate their information of interest. To address these issues, the notion of schema summarization was proposed in the literature [2, 3].

Schema summarization involves identifying semantically related schema elements, representing what users may perceive as a single unit of information in the schema. Identifying abstract representations of schema entities helps in efficient browsing and better understanding of complex database schema. Practical applications of schema summarization are as follows:

- *Schema Matching* [4, 5] is a well researched issue. Schema matching involves identifying mappings between attributes from different schemas. After

S. Srinivasa and V. Bhatnagar (Eds.): BDA 2012, LNCS 7678, pp. 77–90, 2012.

identifying abstract representations of schema elements, we can reduce the number of mapping identification operations by identifying mappings between abstract levels rather than schema level.
- In *Query Interfaces*, users construct their query by selecting tables from schema. A quick schema summary lookup might help the user in understanding where his desired information is located and how is it related to other entities in the schema.

The problem of schema summarization has gained attention recently in the database community. Existing approaches [3, 6, 7] for generating schema summary exploit two main sources of database information, the database schema and data stored in the database. In another related work, Wu et.al. [8] described an elaborate approach (*iDisc*) for clustering schema elements into topical structures by exploiting the schema and the data stored in the database.

In this paper, we propose an alternative approach for schema summarization by exploiting the database documentation, in addition to schema. It can be noted that we investigated how documentation of the database provides the scope for efficient schema summarization. The database documentation contains domain specific information about the database which can be used as an information source. For each table, first we identify the corresponding passages in the documentation. Later, a table similarity metric is defined by exploiting similarity of passages describing the schema elements in the documentation and the referential relationships between tables. Using the similarity metric, a greedy weighted k-center clustering algorithm is used for clustering tables and generating schema summary. The experimental results on TPCE benchmark database shows the effectiveness of the proposed approach.

The rest of the paper is organized as follows; Section 2 discusses related work. In section 3, we describe the proposed approach including the basic idea, table similarity measure and clustering algorithm. In section 4, we discuss the experimental results and analysis. Section 5 includes conclusions and future works.

2   Related Work

In [3], authors proposed summarization algorithms to identify important schema entities while providing broad information coverage in XML schema. However, some of the assumptions made for XML schema summarizations could not be applied for relational schema. Yang et.al [7] proposed an improved algorithm for relational schema summarization. In [6], authors use community detection techniques and table importance measure proposed in [7] to generate schema summary for large scale databases. Wu et.al. [8] proposed an elaborate approach, *iDisc*, which first models the database into different representations (graph based, vector-based, similarity-based) and then combines clustering from each representation using a voting scheme. However, these approaches are data oriented, utilizing schema and data available in the tables. In contrast, the proposed approach uses schema information and database documentation to generate schema summary.

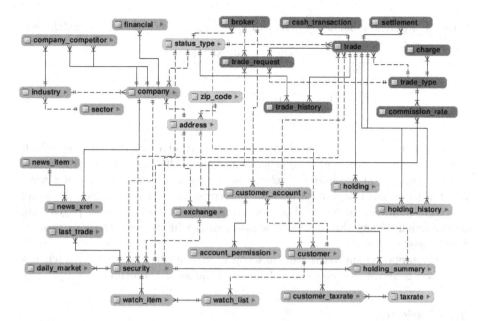

Fig. 1. TPCE Schema

Schema matching [5] involves identifying mappings between schema elements. [9, 10] approach are schema oriented for finding correlated schema elements using name, description, relationship and constraints. In [11] authors have proposed an integrated approach, using linguistic approaches and structure matching process. Identifying mappings is analogous to finding similarity between schema elements belonging to two different schema. However, the proposed approach is concentrated on finding correlated schema elements in the same schema.

3 Proposed Approach

We use the TPCE schema [12] described in Figure 1 as the running example in this paper. The TPCE schema consists of 33 tables which are grouped into four categories of tables: *Customer (blue), Market (green), Broker (red) and dimension (yellow)*. This categorization is provided by the TPC-E benchmark and it also serves as the gold standard for evaluation in our experiments.

Existing approaches for clustering database tables are data oriented, utilizing schema and data in the database for generating schema summary. In scenarios where the data is insufficient or some tables do not contain data, we have to look for alternate sources of information. For example, in the TPCE benchmark database, if no active transactions are considered, the table *trade_request* is empty and hence, cannot be considered for clustering in existing approaches. Thus, we investigate alterative sources of information for a database. Databases are generally accompanied with the *documentation* or the *requirement document*.

Table 1. Referential Similarity between tables security, daily_market and watch_item

	Security	Daily Market	Watch Item
Security	-	1	1
Daily Market	1	-	0
Watch Item	1	0	-

These documents contain domain specific information about the database which could be exploited for generating schema summary. Although one can go through the documentation and infer the schema summary manually, but it is not always feasible to do so. Documentations for enterprise database are generally large, spanning hundreds of pages. The documentation for TPCE is 286 pages long and manually going through the documentation will thus be a tedious process for the user.

In the proposed approach, we aim to propose an efficient approach for schema summary generation, using only schema and the *documentation*.

3.1 Basic Idea

A foreign key relationship between two tables shows that there exists a semantic relationship between two tables. However, referential relationships alone do not provide good results [7]. Hence, we attempt to supplement this referential similarity between tables with another notion of similarity, such that the tables belonging to one category attain higher intra-category similarity. This additional similarity criteria is based on finding similarity between the passage of text representing the table in the database documentation. The intuition behind this notion of similarity is that the tables belonging to the same categories should share some common terms about the category in the *documentation*. We combine the referential similarity and document similarity by means of a weighed function and obtain a table similarity metric over the relational database schema. After pairwise similarity between tables is identified, we use a *Weighted K-Center* clustering algorithm to partition tables into k clusters.

We propose a measure for table similarity. The measure has two components: one based on referential relationship and the other based on similarity of corresponding passages in the documentation. We first explain about the components and then discuss the table similarity measure.

3.2 Schema Based Table Similarity

In a relational database, foreign keys are used to implement referential constraints between two tables. Presence of foreign keys thus implies that the two tables have a semantic relationship. Such constraints are imposed by the database designer or administrator and form the basic ground truth on similarity between tables. In our approach, referential similarity is expressed as *RefSim(R,S)*.

$$RefSim(R, S) = \begin{cases} 1 \text{ , If R,S have foreign key constraint} \\ 0 \text{ , Otherwise.} \end{cases}$$

Example1. Consider the three tables *Security, Daily_market and Watch_item* (*S,D* and *W*) in the TPCE schema. Table *security* has foreign key relationship with *daily_market* and *watch_item*, hence $RefSim(S, D) = RefSim(D, S) = 1$ and $RefSim(S, W) = RefSim(W, S) = 1$. The pairwise similarity is described in Table 1.

3.3 Documentation Based Table Similarity

In addition to the referential similarity we also try to infer similarity between tables using *database documentation* as an external source of information. First, we find the passage describing the table in the documentation using passage retrieval approach. Similarity between two tables thus corresponds to similarity between the corresponding passages in the documentation. The passage from the documentation representing a table T_i is referred to as the table-document of T_i, $TD(T_i)$. The first task is to identify the table-document for each table from the *documentation*. Later, we find pairwise similarity between the table-documents.

Finding Relevant Text from the Documentation. Passage retrieval [13–17] is a well researched domain. Passage retrieval algorithms return the top-m passages that are most likely to be the answer to an input query. We use a sliding window based passage retrieval approach similar to the approach described in [18]. In this paper, we focus on using a passage retrieval approach to evaluate table similarity from database documentation rather than comparing different approaches for passage retrieval from the documentation.

Consider a table T_i with a set of attributes $A_i = (A_{i1}, A_{i2}..A_{ik})$. Given a database documentation (D), for each table T_i we construct a query $Q(T_i)$ consisting of the table name and all its attributes as keywords.

$$Q(T_i) =< T_i, A_{i1}, A_{i2}..A_{ik} > \tag{1}$$

In a sliding window based passage retrieval approach, given a window size w_i for T_i, we search w_i continuous sentences in the document sequentially for the keywords in $Q(T_i)$. If at any instance, the window matches all the keywords from $Q(T_i)$, the passage in the window is considered a potential table-document for T_i. In cases where multiple windows are identified, we implement a ranking function [19] for the retrieved passages and choose the passage with the highest ranking score. The selection of an appropriate window size is a crucial step as the number of keywords in $Q(T_i)$ varies for each T_i. We propose two types of window functions:

- Independent window function, $f(Q(T_i)) = c$
- Linear window function, $f(Q(T_i)) = a \times |Q(T_i)| + c$

After the passage describing the table is identified, we store the passage in a separate document and represent it as the table-document $TD(T_i)$ for the table table T_i.

Similarity of Passages. Once the table-documents have been identified, we have a corpus containing table-document(s) for each table. The table-document(s) are preprocessed by removing stop-words and performing stemming using Porter Stemmer. The table-document can be modeled in two ways:

- **TF-IDF Vector:** $TD(i) = (w_1, w_2, ..w_d)$ can be represented as a d-dimension TFIDF feature vector $d = |corpus|$ and w_i represents the TFIDF score for the ith term in $TD(i)$.
- **Binary Vector:** $TD(i)$ is represented as a d-dimension binary vector $TD(i) = (w_1, w_2, ..w_d)$, where $d = |corpus|$ and w_j is 1 if TD(i) contains the term w_j and 0 otherwise.

We then calculate pairwise similarity between table-documents using the *cosine similarity* measure or the *jaccard coefficient*:

$$DocSim_{cos}(R, S) = DocSim(doc_R, doc_S) = \frac{doc_R.doc_S}{|doc_R| \times |doc_S|} \qquad (2)$$

$$DocSim_{jacc}(R, S) = DocSim(doc_R, doc_S) = \frac{doc_R \cap doc_S}{|doc_R| \cup |doc_S|} \qquad (3)$$

3.4   Table Similarity Measure

For two tables R and S, let $RefSim(R, S)$ represent the referential similarity and $DocSim(R, S)$ represent the document similarity between R and S. We combine the referential similarity and document similarity using a weighing scheme as

$$Sim(R, S) = \alpha \times RefSim(R, S) + (1 - \alpha) \times DocSim(R, S) \qquad (4)$$

Where α is a user specified parameter called the *contribution factor* $0 \leq \alpha \leq 1$. It measures the contribution of referential similarity to the table similarity. In some cases, two tables have a low value of (combined) similarity, but have high similarity to a common neighboring table. For example, in Figure 1, tables *account_permission(AP)* and *customer(C)* do not have a referential similarity but both are similar to the table *customer_account(CA)*. In such cases two tables gain similarity as they have similar neighbors. For the previous example, similarity between *account_permission* and *customer* should be max(Sim(AP,C) , Sim(AP,CA) × Sim(CA,C)).

We construct an undirected database graph $G = (V, E)$, where nodes(V) correspond to tables in the database schema. For any two tables R and S, we define an edge, representing the combined similarity $Sim(R, S)$ between the tables defined in (3). The database graph G is a complete graph.

Algorithm 1. Finding Table Similarity

Input: D: Database Schema, TD: Set of Table-Document vectors, S: Document similarity measure, α: Contribution factor,
Output: Sim: Pairwise similarity between tables in database

$RefSim \leftarrow$ REFERENCE-SIMILARITY(D)
$DocSim \leftarrow$ DOCUMENT-SIMILARITY(TD, S)
$Sim \leftarrow \alpha \times RefSim + (1 - \alpha) \times DocSim$
for all *tables* as k **do**
    for all *tables* as i **do**
        for all *tables* as j **do**
            if $Sim(i, k) \times Sim(k, j) < Sim(i, j)$ **then**
                $Sim(i, j) \leftarrow Sim(i, k) \times Sim(k, j)$
            end if
        end for
    end for
end for
return Sim

procedure REFERENCE-SIMILARITY(D)
    for all *tables* as R **do**
        for all *tables* as S **do**
            if *R,S have foreign key relationship in D* **then**
                $RefSim(R, S) \leftarrow 1$
            else
                $RefSim(R, S) \leftarrow 0$
            end if
        end for
    end for
    return $RefSim$
end procedure

procedure DOCUMENT-SIMILARITY(TD, S)
    for all *tables* as R **do**
        for all *tables* as S **do**
            $DocSim(R, S) \leftarrow S(TD(R), TD(S))$
        end for
    end for
    return $DocSim$
end procedure

Consider a path $p : R = T_i, T_{i+1}, T_{i+2}, ...T_j = S$ between two tables T_i and T_j. Similarity between the tables T_i and T_j along path p is

$$Sim_p(R, S) = \prod_{k=i}^{j-1} Sim(T_k, T_{k+1}) \qquad (5)$$

Then the path with the maximum similarity between R and S gives the complete similarity between R and S.

$$Sim(R, S) = max_p Sim_p(R, S) \qquad (6)$$

As we construct a complete graph, we use the Floyd-Warshall algorithm for finding the shortest paths in a weighted graph. In our case, we define the shortest distance as having the maximum similarity. Since we construct a complete graph for finding all pairs maximum similarity paths, the algorithm takes $\mathcal{O}(n^3)$ running time for this step.

Algorithm 1 describes the procedure for calculating the pairwise similarity between tables in a schema. By taking the database schema, set of extracted passages, a document similarity measure and contribution factor as input, the algorithm return pairwise similarity between tables. First we calculate the referential and document similarity for $\mathcal{O}(n^2)$ pairs and later combine them using the contribution factor. The procedure reference-similarity() takes as input the database schema and calculates the similarity between two tables based on the referential relationships. The procedure document-similarity() takes as input the passage corresponding to each table, a document similarity measure and calculates similarity between tables based on the similarity of corresponding passages of the tables. Note that for every table, the passage is extracted by employing the passage retrieval approach described in section 3.3.

3.5 Clustering Algorithm

For generating summary, we use a greedy Weighted K-Center clustering algorithm. It provides a min-max optimization problem, where we want to minimize the maximum distance between a table and its cluster center.

Influential Tables and Cluster Centers. In schema summarization, the notion of *influential* table is used for clustering [7]. The notion says that the most important tables should not be grouped in the same cluster. We measure the influence of a table by measuring the influence one table has on other tables in the schema [20]. Specifically, if a table is closely related to large number of tables in the database, it will have a high influence score. The influence score helps in identifying the cluster centers, described in the clustering process. The influence of a table R on another table S in the database schema is defined as

$$f(R, S) = 1 - e^{-Sim(R,S)^2} \qquad (7)$$

Influence score of a table is thus defined as

$$f(R) = \sum_{t_i \in T} f(R, t_i) \qquad (8)$$

where T represents the set of tables in the database.

Clustering Objective Function. The clustering objective function aims to minimize the following measure [7].

$$Q = max_{i=1}^{k} max_{R \epsilon C_i} f(R) \times (1 - Sim(R, Center(C_i))) \qquad (9)$$

where k in the number of clusters, $f(R)$ is the influence score of table R and $Center(C_i)$ represents the center of the ith cluster (C_i).

Clustering Process. We use the weighted k-center algorithm which considers the influence score for clustering. In this approach, the most influential table is selected as the first cluster center and all the tables are assigned to this cluster. In each subsequent iterations, the table with lowest weighted similarity from its cluster center separates out to form a new cluster center. The remaining tables are re-assigned to the closest cluster center. We repeat the process for k iterations, such that k clusters are identified for the database schema. The time complexity of the greedy clustering algorithm is $\widetilde{\mathcal{O}}(kn^2)$ [21].

4 Experimental Results

In this section, we present results of experiments conducted on our proposed approach. The following parameters have been used at different stages in our approach:

- Window size function (f) for table-document discovery.
- Document similarity measure (S), for calculating similarity of passage describing the tables in the documentation.
- α, the contribution factor in combined table similarity metric.
- k, the number of clusters determined by the clustering algorithm.

Varying any one of the parameters affects the table similarity metric and clustering. We study the influence of these parameters by varying one parameter while keeping the other parameters constant. Later, we conduct experiments on the clustering algorithm and compare our approach with other existing approaches.

4.1 Experimental Setup

We used the TPCE database[12], provided by TPC. It is an online transaction processing workload, simulating the OLTP workload of a brokerage firm. TPC also provides a software package *EGen* to facilitate the implementation of the TPCE database. We used the following parameters to implement an instance of TPCE: Number of Customers = 5000, Scale factor = 36000, Initial trade days = 10.

The TPCE schema consists of 33 tables, which are grouped into four categories: *Customer, Market, Broker and Dimension*. We use this categorization as the gold standard to measure the accuracy of our approach. The dimension tables are not an explicit category, they are used as companion tables to other fact tables and hence can be considered as outliers to our clustering process. We thus aim to cluster the other 29 tables any measure the accuracy of these 29 tables to the given gold standard.

In addition, TPCE also provides the documentation for the TPCE benchmark. It is a 286 page long document and contains information about TPCE buisness and application environment, the database and the database transactions involved. This document serves as an external source of information as described in the proposed schema summarization approach.

4.2 Evaluation Metric

The accuracy of clustering and table similarity metric is evaluated by means of an *accuracy* score, proposed in [7]. The *accuracy* score has different connotations for clustering evaluation and table similarity evaluation. For the table similarity metric, we find the top-n neighbors for each table based on the *Sim* metric described in Equation (6). Unless specifically mentioned, we find the top-5 neighbors in our experiments. From the goldstandard, if category of table T_i is C_a, m_i is the count of the tables in the top-n neighborhood of T_i belonging to the same category as C_a, then average accuracy of similarity metric is defined as

$$acc_{sim} = \frac{\sum_{i \epsilon T} \frac{m_i}{n}}{|T|} \qquad (10)$$

Similarly for clustering accuracy, consider a cluster i containing n_i number of tables. If the category of the cluster center of a cluster i is C_a; let m_i denote the count of tables in the cluster that belong to the category C_a. Then accuracy of the cluster i and overall clustering accuracy is

$$acc_{clust_i} = \frac{m_i}{n_i} \qquad (11)$$

$$acc_{clust} = \frac{\sum_{i \epsilon T} m_i}{|T|} \qquad (12)$$

4.3 Effect of Window Function (f) on Combined Table Similarity and Clustering

In this experiment we measure the impact of varying the window function f for window size (w) on the clustering accuracy and table similarity metric. We fix $\alpha = 0.5$, $k = 3$ and use the tf-idf based cosine similarity for table-document similarity. We conduct an experiment with the following window functions

- $w_i = f(Q(T_i)) = 10$
- $w_i = f(Q(T_i)) = 20$
- $w_i = f(Q(T_i)) = 2 \times |Q(T_i)| + 1$
- $w_i = f(Q(T_i)) = 3 \times |Q(T_i)| + 1$

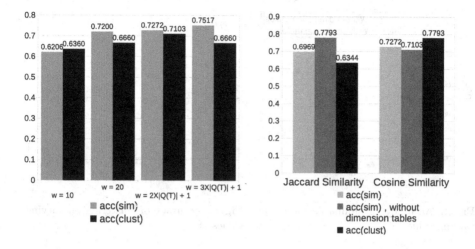

Fig. 2. acc_{sim} and acc_{clust} values on varying window function, f

Fig. 3. acc_{sim} and acc_{clust} values for document similarity functions S

The results of this experiments are shown in Figure 2. We observe that although the function $f = 20$ gives respectable results, it is hard to determine a value of such constant($f = 10$ gives poor results). Using a constant window size can cause loss of information in some cases or add noise in other cases. To be on the safe side, linear window functions, which gave comparatively similar results are preferred. In further experiments, unless specified specifically, we use the window function as $f(Q(T_i)) = 2 \times |Q(T_i)| + 1$.

4.4   Effect of Document Similarity Measure (S) on Similarity Metric and Clustering Accuracy

The table-documents identified for each table can be of variable length. We study two similarity measures described in Equation (2) and Equation (3), cosine similarity and jaccard similarity. We compare the accuracy of similarity metric and clustering algorithm for the two similarity measures for $k = 3$, $\alpha = 0.5$ and $f(Q(T_i)) = 2 \times |Q(T_i)| + 1$. The results of the experiments are shown in Figure 3. We observe that the tf-idf score based Cosine similarity measure is consistent in the results. This can be attributed to the fact the table-documents share a lot of similar terms about the domain of the document and hence term frequency and inverse document frequency play an important role in determining the score of the terms in a document.

4.5   Effect of Contribution Factor (α) on Table Similarity and Clustering

In this section we measure the impact of varying α on the clustering accuracy and table similarity metric. In this experiment, we fix $w = 2 \times |Q(T_i)|$ and $k =$

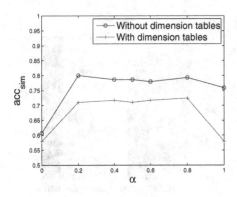

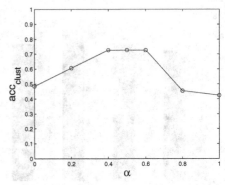

Fig. 4. Accuracy of similarity metric on varying values of α

Fig. 5. Accuracy of clustering on varying values of α

3, while varying α from 0 to 1. Figure 4 and Figure 5 show the results of varying α on clustering accuracy and accuracy of similarity metric. One interesting observation is we achieve best clustering accuracy when the contribution of referential similarity and document similarity are almost equal ($\alpha = 0.4, 0.5, 0.6$). This shows that rather than one notion of similarity supplementing the other, both similarities have equal importance in generating schema summary. Also using any single similarity measure (when α is 0 or 1) produces low accuracy results which verifies the claims made in this paper.

4.6   Comparison of Clustering Algorithm

In this section, we compare clustering algorithms for schema summary. In addition to the proposed weighted k-center clustering algorithm using an influence function($Clust_s$), we implement the following clustering algorithms:

- $Clust_c$, A community detection based schema summarization approach proposed in [6].
- $Clust_d$, the schema summarization approach proposed in [7]. The clustering approach uses a table importance metric based weighted k-center clustering algorithm.
- $Clust_v$, Combines results from clustering using reference similarity and clustering using document similarity using a voting scheme similar to [8]. This algorithm focuses on combining clustering from different similarity models rather than combining similarity models.

Figure 6 shows the clustering accuracy achieved for $k = (2, 3, 4)$ for various clustering algorithms. We observed that $clust_s$ and $clust_d$ achieve almost similar accuracy, with $clust_s$ giving slightly higher accuracy as it was able to successfully cluster the table *trade_request*. If no active transactions are considered for the TPCE database, the table *trade_request* is empty and data oriented approaches

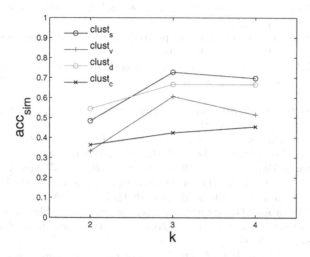

Fig. 6. Clustering accuracy for different clustering algorithms

are unable to classify the table. For the $clust_v$ and $clust_c$ approaches, no specific patterns were observed. The reason for low accuracy of $clust_v$ is because referential similarity provides a very imbalanced and ineffective clustering which deters the overall clustering accuracy in the voting scheme significantly.

5   Conclusions and Future Work

Schema summarization has been proposed in the literature to help users in exploring complex database schema. Existing approaches for schema summarization are data-oriented. In this paper, we proposed a schema summarization approach for relational databases using database schema and the database documentation. We proposed a combined similarity measure to incorporate similarities from both sources and proposed a framework for summary generation. Experiments were conducted on a benchmark database and the results showed that the proposed approach is as effective as the existing data oriented approaches. We plan to extend this work by developing algorithms for learning the values of various parameters used in the proposed approach.

References

1. Nandi, A., Jagadish, H.V.: Guided interaction: Rethinking the query-result paradigm. PVLDB 4(12), 1466–1469 (2011)
2. Jagadish, H.V., Chapman, A., Elkiss, A., Jayapandian, M., Li, Y., Nandi, A., Yu, C.: Making database systems usable. In: Proceedings of the 2007 ACM SIGMOD International Conference on Management of Data, SIGMOD 2007, pp. 13–24. ACM, New York (2007)
3. Yu, C., Jagadish, H.V.: Schema summarization, pp. 319–330 (2006)

4. Doan, A., Halevy, A.Y.: Semantic-integration research in the database community. AI Mag. 26(1), 83–94 (2005)
5. Rahm, E., Bernstein, P.A.: A survey of approaches to automatic schema matching. The VLDB Journal 10(4), 334–350 (2001)
6. Xue Wang, X.Z., Wang, S.: Summarizing large-scale database schema using community detection. Journal of Computer Science and Technology, SIGMOD 2008 (2012)
7. Yang, X., Procopiuc, C.M., Srivastava, D.: Summarizing relational databases. Proc. VLDB Endow. 2(1), 634–645 (2009)
8. Wu, W., Reinwald, B., Sismanis, Y., Manjrekar, R.: Discovering topical structures of databases. In: Proceedings of the 2008 ACM SIGMOD International Conference on Management of Data, SIGMOD 2008, pp. 1019–1030. ACM, New York (2008)
9. Bergamaschi, S., Castano, S., Vincini, M.: Semantic integration of semistructured and structured data sources. SIGMOD Rec. 28(1), 54–59 (1999)
10. Palopoli, L., Terracina, G., Ursino, D.: Experiences using dike, a system for supporting cooperative information system and data warehouse design. Inf. Syst. 28(7), 835–865 (2003)
11. Madhavan, J., Bernstein, P.A., Rahm, E.: Generic schema matching with cupid. In: Proceedings of the 27th International Conference on Very Large Data Bases, VLDB 2001, pp. 49–58. Morgan Kaufmann Publishers Inc., San Francisco (2001)
12. TPCE, http://www.tpc.org/tpce/
13. Clarke, C.L.A., Cormack, G.V., Kisman, D.I.E., Lynam, T.R.: Question answering by passage selection (multitext experiments for trec-9). In: TREC (2000)
14. Ittycheriah, A., Franz, M., Jing Zhu, W., Ratnaparkhi, A., Mammone, R.J.: Ibm's statistical question answering system. In: Proceedings of the Tenth Text Retrieval Conference, TREC (2000)
15. Salton, G., Allan, J., Buckley, C.: Approaches to passage retrieval in full text information systems. In: Proceedings of the 16th Annual International ACM SIGIR Conference on Research and Development in Information Retrieval, SIGIR 1993, pp. 49–58. ACM, New York (1993)
16. Tellex, S., Katz, B., Lin, J., Fernandes, A., Marton, G.: Quantitative evaluation of passage retrieval algorithms for question answering. In: Proceedings of the 26th Annual International ACM SIGIR Conference on Research and Development in Informaion Retrieval, SIGIR 2003, pp. 41–47. ACM, New York (2003)
17. Wang, M., Si, L.: Discriminative probabilistic models for passage based retrieval. In: Proceedings of the 31st Annual International ACM SIGIR Conference on Research and Development in Information Retrieval, SIGIR 2008, pp. 419–426. ACM, New York (2008)
18. Xi, W., Xu-Rong, R., Khoo, C.S.G., Lim, E.-P.: Incorporating window-based passage-level evidence in document retrieval. JIS 27(2), 73–80 (2001)
19. Robertson, S., Walker, S., Jones, S., Hancock-Beaulieu, M., Gatford, M.: Okapi at trec-3, pp. 109–126 (1996)
20. Zhou, Y., Cheng, H., Yu, J.X.: Graph clustering based on structural/attribute similarities. Proc. VLDB Endow. 2(1), 718–729 (2009)
21. Dyer, M., Frieze, A.: A simple heuristic for the p-centre problem. Oper. Res. Lett. 3(6), 285–288 (1985)
22. Rangrej, A., Kulkarni, S., Tendulkar, A.V.: Comparative study of clustering techniques for short text documents. In: Proceedings of the 20th International Conference Companion on World Wide Web, WWW 2011, pp. 111–112. ACM, New York (2011)

Faceted Browsing over Social Media

Ullas Nambiar[1], Tanveer Faruquie[2], Shamanth Kumar[3], Fred Morstatter[3],
and Huan Liu[3]

[1] EMC India COE, Bangalore, India
Ullas.Nambiar@emc.com
[2] IBM Research Lab, New Delhi, India
ftanveer@in.ibm.com
[3] Computer Science & Engg, SCIDSE, Arizona State University, USA
{skumar34,fmorstat,huan.liu}@asu.edu

Abstract. The popularity of social media as a medium for sharing information has made extracting information of interest a challenge. In this work we provide a system that can return posts published on social media covering various aspects of a concept being searched. We present a faceted model for navigating social media that provides a consistent, usable and domain-agnostic method for extracting information from social media. We present a set of domain independent facets and empirically prove the feasibility of mapping social media content to the facets we chose. Next, we show how we can map these facets to social media sites, living documents that change periodically to topics that capture the semantics expressed in them. This mapping is used as a graph to compute the various facets of interest to us. We learn a profile of the content creator, enable content to be mapped to semantic concepts for easy navigation and detect similarity among sites to either suggest similar pages or determine pages that express different views.

1 Introduction

Social media is defined as *a group of Internet-based applications that connects humans via Web and mobile phone technologies for the purpose of allowing the creation and exchange of user generated content*. Social media is represented in various forms such as magazines, Internet forums, weblogs, social blogs, microblogging, wikis, podcasts, etc. The number of users, and as a result, the amount of data on social media are ever-increasing.

Due to the similarity of social media content pages with web pages, the first ideas for modeling social web were derived from Web graph models. However, the rich link structure assumption does not hold true in the social web. The sparseness of the hyperlink structure in blogs was shown in [1]. Furthermore, the level of interaction in terms of comments and replies to an entry made on social media page makes the the social web different from the web. Figure 1 shows a graphical visualization of the network of blogs that we constructed using a dataset consisting of nearly 300 blogs from the BLOG06 test data collection in TREC06 [1]. The nodes in the graph represent individual

---

[1] http://trec.nist.gov

S. Srinivasa and V. Bhatnagar (Eds.): BDA 2012, LNCS 7678, pp. 91–100, 2012.
© Springer-Verlag Berlin Heidelberg 2012

blogs and an edge is drawn between nodes if the blogs show some similarity. The similarity was estimated using the Cosine Similarity metric after the removal of stopwords and computing TF-IDF weights for key terms. Contrary to our expectations, the graph turned out to be highly connected even with the similarity threshold set to a very high 0.8 with 1.0 suggesting an exact replica. We can clearly see groups of blogs coming together and forming local clusters.

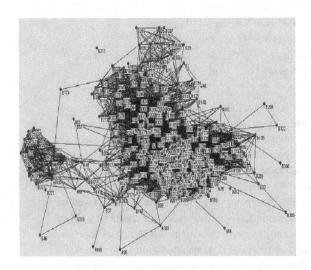

Fig. 1. Similarity Graph for Blogs with Threshold > 0.8

Developing an alternative to keyword search is a challenge. A close variant of keyword search is the *Faceted search or browsing* technique where information is organized into multiple orthogonal features called facets [2]. The *colon classification* [3] scheme was developed by S. R. Ranganathan, to apply faceted classification to the physical world, specifically for the purpose of organizing books. Specifically, he proposed five facets *Personality, Matter, Energy, Space* and *Time* for classifying most of the physical world. In this model Personality stood for a person or event being classified, Matter for what it is composed of and Energy for depicting the change or evolution. Space and Time are self-explanatory.

We believe that Ranganathan's classic PMEST model would be enough to provide an easy to understand, widely acceptable faceted browsing system for social media users. Hence, we propose to mine and use five distinct and orthogonal facets to classify a collection of data originating from social media. These are the *profile* of the person authoring the content, the *topics* it deals with and the *purpose* for which the content is written. These three facets loosely map to Personality, Matter and Energy in the PMEST model. The Space and Time facets can be mapped to geography(location) and date. This information is often appended by the social media service providers. Furthermore, both these dimensions are well understood by all users and hence form an easy to classify and navigate facet that is domain independent.

In summary, this paper describes our vision of how the social media should be presented to a user on the Web. The main contributions of our paper are:
– We initiate research into the problem of automated, semantics driven and domain independent faceted classification for social media.

– We propose five facets for describing any Social Media dataset based on the classical PMEST classification scheme from Library Sciences.
– We present methods to automatically and efficiently extract the proposed facets and present a end-to-end system built to showcase the usability of the facets.

2 Related Work

The blogosphere comprises of several focused groups or communities that can be treated as sub-graphs. These communities are highly dynamic in nature that have fascinated researchers to study its structural and temporal characteristics. Kumar et. al. [4] extended the idea of hubs and authorities and included co-citations as a way to extract all communities on the web and used graph theoretic algorithms to identify all instances of graph structures that reflect community characteristics. Another effort in this area has been on the analysis of information flow within these link networks [5]. Structure of blog graphs has been exploited for discovering social characteristics like identifying influential bloggers [6], discovering blogging patterns in the blogosphere [7] and community factorization using the structural and temporal patterns of the communications between bloggers [8].

Often microblogging sites allow their users to provide tags to describe the entry. Tags are also used to discriminate between pages in the social network [9]. The drawback of tagging is that it is difficult to associate a "meaning", induce a hierarchy or relate different tags. Brooks and Montanez [10] presented a study where the human labeled tags are good for classifying blog posts into broad categories while they were less effective in indicating the particular content of a blog post. In another research [11], authors tried to cluster blog posts by assigning different weights to title, body and comments of a blog post. However, these approaches rely on the keyword-based clustering which suffers from high-dimensionality and sparsity.

Various forms of community-aware ranking methods have been developed, mostly inspired by the well-known PageRank method [12] for web link analysis. [13] proposes FolkRank for identifying important users, data items, and tags. SocialPageRank to measure page authority based on its annotations and SocialSimRank for the similarity of tags is introduced in [14].

3 Faceted Browsing over Social Media

In this section we share technical details and challenges involved in mapping the social media datasets to the five facets we chose based on the PMEST classification scheme. As shown in Figure 2, given any type of social media dataset, we extract the content using the extractors and APIs available in the *Content Capture* module. This module is also responsible for removing unwanted metadata and links from content pages. We clean the extracted information by removing unnecessary tags from the post-body. The next module *Feature Engineering* uses a series of steps aimed at *Cleaning* the content by running *word stemmers*. Then we remove frequently occurring terms (*stop word elimination*) and *N-gram extraction* as these are well-documented methods used by most document processing systems. We consider each social media page as a document with some uniform characteristics - namely, the presence of a *permalink, post id, unique identifier* and the *post-body*.

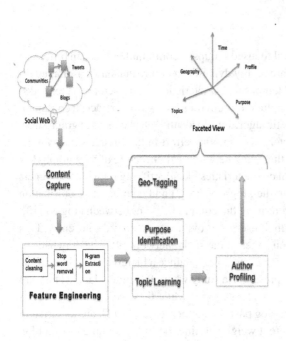

Fig. 2. Flowgraph of our System

Once the features are extracted, we use the cleaned content to perform *Topic Learning, Purpose Identification, Geo-Tagging* and and *Profile creation*. We believe the techniques used for extracting *topics* can be used to also learn the *purpose* with the appropriate terms being used to model the learning process.

3.1   Topic Extraction

Topic extraction is a well studied field and several methods have been proposed to extract topics from documents. Since we are not making any assumptions about the domain, knowledge of writers, or the interests of the users we use a completely unsupervised data driven approach to automatically extract topics from blogs.

Topics Using LDA. In particular, we use the Latent Dirichlet Allocation (LDA) model [15], that considers topics to be multinomial probability distribution over the vocabulary of all the blogs under consideration. This model has the flexibility that the topics can be learned in a completely unsupervised manner.

If there are D blogs under consideration such that d^{th} blog has N_d words represented as w_d, picked from a vocabulary of size V, and the total number of topics talked by bloggers is K then the generative process is as follows

1. For each topic $k \in \{1 \dots K\}$ choose a topic as a V dimensional multinomial distribution ϕ_k over V words. This is from a V dimensional Dirichlet distribution β
2. For each blog $d \in \{1 \dots D\}$ having N_d words
   (a) Choose the topics talked about in that blog as a K-dimensional multinomial θ_d over the K topics. This is from a K-dimensional Dirichlet distribution α
   (b) For each word position $j \in \{1 \dots N_d\}$ in blog d

i. Select the topic $z_{jd} \in \{1 \dots K\}$ for the position from the multinomial θ_d

ii. Select the word w_{jd} for this topic drawn from the multinomial ϕz_{jd}

In the above process ϕ and θ are unknown parameters of the model. We use Gibbs sampling [16] to find these parameters. Since Dirichlet is the conjugate prior of the multinomial distribution we can collapse the parameters θ and ϕ and only sample the topics assignments z.

Topics Based on Word Clouds. A word cloud is a visual representation of words, where the importance of each word is represented by its font size with the most important words the largest. Here each word is representative of a topic. An entry (such as blog post or tweet) t can be decomposed into its constituent words (unigrams). One way to determine the importance of each word t_w in the tweet is by computing its TF-IDF score. The importance score of t_w is

$$\mathcal{I}(t_w) = \frac{N_w}{N} \times \log \frac{|T|}{|T_w|}, \tag{1}$$

where N_w is the number of times a word w has occurred in T, N is the vocabulary size of the corpus, $|T| = M$, and $|T_w|$ is the count of the number of entries that contain the word w. The first part of the product is the term frequency score(TF) of the word and the second part is the inverse document frequency score(IDF). Given a desired maximum font size of F_{max} and the $\mathcal{I}(t_w)$, the font size f_{t_w} of word t_w is

$$f_{t_w} = \lfloor F_{max} \times \mathcal{I}(t_w) \rfloor. \tag{2}$$

Computation of IDF requires too many queries to the underlying storage system resulting in high I/O cost. In our system, we are more concerned about identifying frequent words that can help us convey the topic expressed in the entries so we do not compute IDF scores and simply use TF after the removal of stop words.

3.2   Extracting Author Profiles from Social Media

One of the dimensions along which we want to enable the social media search is the profile of the author creating the content. Our interest is in identifying profiles or roles that can span many domains e.g. expert/specialist, generalist, self-blogger, observer/commentator, critique, etc. The easiest to identify is whether a person is an *expert* in a given area or if she has an opinion on everything - *a generalist* . We believe experts would have less noise or randomness in their posts. Entropy [17] estimation is used in Information Theory to judge the disorder in a system. Shannon's entropy is defined as follows. Let X be a discrete random variable on a finite set $\mathcal{X} = \{x_1, \dots, x_n\}$, with probability distribution function $p(x) = \Pr(X = x)$. The entropy $H(X)$ of X is

$$H(X) = - \sum_{x \in \mathcal{X}} p(x) \log_b p(x). \tag{3}$$

Table 1. Topic Labels learned from Wiki Categories

Topic	Keywords	Mid-Sup	Max-Sup
1	just, christmas, fun, home	Band	Album
2	war, being, human, actor	Book	Actor
3	things, looking, world	Supernatural	Place
4	rss, search, score, news	Single	Place
5	users, pretty, little	Book	Place
6	people, police, year, canada	Company	Place
7	university, body, quizfarm	Bird of Suriname	Place
8	free, body, news, pictures	Place	Place
9	need, home, work, hand	Military Conflict	Vehicle
10	charitycam, kitty, mood, friends	Political Entity	Album

In our case, the system of interest is the social media page or document. For a blog or a twitter account, one can assume that as the number of posts or tweets increase, the disorder and hence the entropy would increase, if the author did not devote himself to commenting on a small subset of topics. The set of topics that we extracted as described above is the finite set X that we use. For each blog post or tweet we then find the probability of mapping it to one of the topic $x \in X$ by counting the number of keywords in the post or tweet which fall under x. Thus, by using entropy we can identify bloggers as *specializing* in a few concepts or as *generalists* who discuss varied concepts.

3.3   Mapping Location and Date

The task of detecting and mapping the *Location* and *Date* of the social media entry felt straight forward to us. Most social content authoring platforms capture the location and the timestamp at which the content was submitted to the system. We found that often, the content itself might be talking about events that happened at a location that was distant from where the author was posting the content. The time of incident described could also vary. Capturing these and using them in mapping requires advanced NLP techniques. We are building annotators to extract and use such information. In this paper, we will not go into the details of those methods.

4   Empirical Evaluation

In this section we provide details of our effort in building a five-faceted social media search system. The rest of the paper will describe our efforts in this direction.

4.1   Datasets Used

We will explain our approach with empirical evidence on two real-world datasets that consist of two very different types of social media - *blogs* and *microblogs*.

Blogs Dataset. We used the BLOG06 test data collection used in the TREC06 [2] Blog track to perform preliminary evaluations and test our hypothesis. This set was created

---

[2] http://trec.nist.gov

Table 2. Labels Assigned by Human Expert to Blog Clusters

No	Topic	Keywords	Cluster
1	Gambling	poker, casino, gamble	2
2	Birthdays	birthday, whatdoesyourbirthdatemeanquiz	3
3	Classifieds	classifieds, coaster, rollerblade, honda	4
4	Podcast	podcasts, radio, scifitalk	5
5	Christmas	christmas, santa, tree, gomer	6
6	Finance	cash, amazon, payday	7
7	Trash	dumb, rated, adult	8,10
8	Comics	comic, seishiro	9
9	Thanksgiving	thankful, food, dinner, shopping	11
10	Movies	movie, harry, potter, rent	13
11	IT	code, project, server, release, wordpress	16
12	Iraq War	war, iraq, media, bush, american	17
13	Cars	car, drive, money, kids	18
14	Books	book, art, science, donor	19
15	Others	dating, cards, nba, napolean, halloween, homework, hair	1, 12, 14, 15, 20

Table 3. Hashtags Used to Crawl Tweet Dataset

Purpose	Hashtags
Occupy Wall Street	#occupywallstreet, #ows, #occupyboston, #occupywallst, #occupytogether, #teaparty, #nypd, #occupydc, #occupyla, #usdor, #occupyboston, #occupysf, #solidarity, #citizenradio, #gop, #sep17, #occupychicago, #15o

and distributed by the University of Glasgow. It contains a crawl of blog feeds, and associated permalink [3] and homepage documents (from late 2005 and early 2006). We took 10,000 permalink documents for our study. We extracted 339 distinct blogs from these documents. We use a small dataset for ease of manually computing the concepts and also verifying the identified relationships.

Tweets Dataset. Using an existing Twitter event monitoring system, TweetTracker [18], we collected tweets discussing the Occupy Wall Street (OWS) movement. The tweets were crawled over the course of 4 months starting from September 14, 2011 to December 13, 2011. A full list of the parameters used is presented in Table 3. The dataset consists of 8,292,822 tweets generated by 852,240 unique users. As in the case of blogs, we identify a smaller set of tweets whose topic can be determined and evaluated manually. These tweets are generated when the number of tweets exceeded the average tweet traffic (102,380) by one standard deviation (77,287). More information on how these days are identified is presented later in the paper. In total there were 10 such days in our dataset which spans 90 days of the OWS movement.

4.2 Topic Extraction

For clustering the blogs extracted from the TREC dataset, we used an LDA implementation where the number of concepts or topics was set to 20. We presented the top 15

---

[3] A permalink, or permanent link, is a URL that points to a specific blog or forum entry after it has passed from the front page to the archives.

words from each distribution of cluster to a small group of human evaluators and asked them to manually label the clusters to real-world concepts. We picked the most common topic as the single best label for a cluster. A few clusters to which no clear semantic label could be assigned was mapped to a generic concept called *Other*.

4.3   Learning Profiles

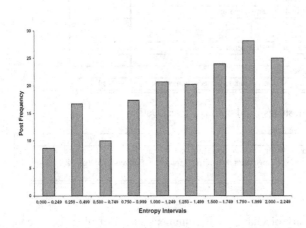

Figure 3 plots the average number of posts for blogs having entropy within a given interval. We can see that blogs with a relatively small number of posts can also have high entropy while blogs with a large number of posts can have low entropy values. We have manually verified that blogs with high post frequency but low entropy map show high weights for a few concepts in our list while blogs with high entropy map to several concepts but with lower strengths. Bloggers who are devoted to a few concepts can be classified as specialists and the bloggers whose blogs show high entropy can be classified as generalists. However, the classification task becomes difficult if the blogger has only a few posts for a concept.

Fig. 3. Entropy of Blogs

4.4   Mapping Tweets to Topics

As our Twitter dataset is related to the *Occupy Wall Street* movement, the *Purpose* of the dataset was pre-determined and hence we did not focus on learning the same. We utilize the Gibbs sampler LDA to discover a coherent topic from the tweets from each selected day. The top 15 words describing these topics are presented in Table 4. By analyzing these topics we find a clear indication of the significant events associated with the OWS movement. For example, around 100 protesters were arrested by the police on October 11th in Boston and the prominent keywords: "boston, police, occupy, wall, street" from the topic for Oct 11 indicate this clearly. To check the veracity of the events extracted, we have compared them to the human edited *OWS Timeline* published in Wikipedia.

Tweet Dataset Mapping with OWS Timeline. To ensure that our approach accurately presents relevant dates to the user, we compare the important dates generated by our method with those known to be distinguished dates in the Occupy movement. Our

Table 4. Topics identified using LDA for the important days

Date	Topic
Oct 11	boston, police, occupy, wall, street, people, protesters, movement, protest, news, obama, media, jobs, 99, arrested
Oct 13	wall, occupy, street, protesters, mayor, bloomberg, park, eviction, defend, tomorrow, people, support, zuccotti, call, police
Oct 14	wall, street, occupy, park, people, protesters, police, movement, tomorrow, global, live, support, today, news, 99
Oct 15	square, times, people, police, occupy, protesters, street, wall, world, nypd, live, movement, arrested, today, sq
Oct 16	occupy, arrested, wall, people, street, protesters, movement, police, protests, protest, obama, support, west, world, times
Oct 17	occupy, wall, street, people, movement, obama, protesters, support, police, protests, protest, world, 99, nypd, party
Oct 18	occupy, wall, street, people, obama, protesters, movement, debate, cain, support, protest, 99, romney, protests, party
Oct 26	police, oakland, occupy, protesters, people, street, march, gas, wall, tear, tonight, movement, live, support, cops
Nov 15	park, police, nypd, protesters, zuccotti, occupy, press, nyc, people, live, street, wall, eviction, bloomberg, raid
Nov 30	lapd, police, live, free, reading, psychic, media, people, protesters, occupy, arrested, city, cops, calling, eviction

Table 5. List of dates with tweets exceeding one standard deviation and the Wikipedia article's explanation of the day's events. Average daily tweets: 102,380.52, Standard Deviation: 77,287.02.

Date	#Tweets	Wikipedia Justification
2011-10-11	186,816	Date not mentioned in article.
2011-10-13	200,835	Mayor Bloomberg tells protesters to leave Zuccotti Park so that it can be cleaned.
2011-10-14	228,084	Zuccotti Park cleaning postponed.
2011-10-15	376,660	Protesters march on military recruitment office to protest military spending.
2011-10-16	194,421	President Obama issues support for the OWS movement.
2011-10-17	193,332	Journalist fired for supporting OWS movement.
2011-10-18	185,743	Date not mentioned in article.
2011-10-26	220,571	Scott Olsen, Military veteran and OWS protester, hospitalized by police at OWS event.
2011-11-15	488,439	NYPD attempts to clear Zuccotti Park.
2011-11-30	203,846	Police arrest protesters in Occupy Los Angeles.

method defines an important date as one in which the number of tweets generated on that day exceeds the average number of tweets generated per day for all of the dates considered for the movement plus one standard deviation. To generate a list of important dates in the movement to which we will compare our method, we scraped all of the dates mentioned in the Wikipedia article "Timeline of Occupy Wall Street"[4]. Table 5 shows the dates we identified as important alongside Wikipedia's explanation for the importance of that date. In all but two instances, the dates we discovered match a landmark date discussed in the Wikipedia article.

5 Conclusion

Social media is a platform with explosive growth in both users and content that can be harnessed to better understand social mores and other aspects of human behavior. In this paper we tackled the problem of enabling easy retrieval of information from the social media by providing a domain agnostic faceted search interface. We presented five

---

[4] http://tinyurl.com/7g5q5st

domain-independent facets that can be extracted from social media content. Next, we offered techniques that can be used to learn these facets. Finally, we demonstrate a system we created to learn these facets with high accuracy, which presents this information in an easily understandable format to its users.

Acknowledgments. This work was sponsored in part by the Office of Naval Research grant: N000141010091.

References

1. Kritikopoulos, A., Sideri, M., Varlamis, I.: Blogrank: Ranking weblogs based on connectivity and similarity features. In: 2nd International Workshop on Advanced Architectures and Algorithms for Internet Delivery and Applications, NY, USA (2006)
2. English, J., Hearst, M., Sinha, R., Swearingen, K., Yee, P.: Hierarchical faceted metadata in site search interfaces. In: CHI Conference Companion (2002)
3. Ranganathan, S.: Elements of library classification. Asia Publishing House (1962)
4. Kumar, R., Raghavan, P., Rajagopalan, S., Tomkins, A.: Trawling the web for emerging cyber communities. In: WWW (1999)
5. Glance, N., Hurst, M., Nigam, K., Siegler, M., Stockton, R., Tomokiyo, T.: Deriving marketing intelligence from online discussion. In: KDD, pp. 419–428. ACM (2005)
6. Agarwal, N., Liu, H., Tang, L., Yu, P.S.: Identifying the influential bloggers in a community. In: WSDM, pp. 207–218. ACM, New York (2008)
7. Leskovec, J., McGlohon, M., Faloutsos, C., Glance, N.S., Hurst, M.: Patterns of cascading behavior in large blog graphs. In: SDM (2007)
8. Chi, Y., Zhu, S., Song, X., Tatemura, J., Tseng, B.L.: Structural and temporal analysis of the blogosphere through community factorization. In: KDD, pp. 163–172. ACM (2007)
9. Qu, L., Müller, C., Gurevych, I.: Using tag semantic network for keyphrase extraction in blogs. In: CIKM, pp. 1381–1382. ACM, New York (2008)
10. Brooks, C.H., Montanez, N.: Improved annotation of the blogosphere via autotagging and hierarchical clustering. In: WWW (2006)
11. Li, B., Xu, S., Zhang, J.: Enhancing clustering blog documents by utilizing author/reader comments. In: Proceedings of the 45th Annual ACM Southeast Regional Conference (2007)
12. Brin, S., Page, L.: The anatomy of a large-scale hypertextual web search engine. Computer Networks (1998)
13. Hotho, A., Jäschke, R., Schmitz, C., Stumme, G.: Information Retrieval in Folksonomies: Search and Ranking. In: Sure, Y., Domingue, J. (eds.) ESWC 2006. LNCS, vol. 4011, pp. 411–426. Springer, Heidelberg (2006)
14. Bao, S., Xue, G., Wu, X., Yu, Y., Fei, B., Su, Z.: Optimizing web search using social annotations. In: WWW, pp. 501–510 (2007)
15. Blei, D., Ng, A., Jordan, M.: Latent dirichlet allocation. Journal of Machine Learning Research 3(4-5), 993–1022 (2003)
16. Porteous, I., Newman, D., Alexander, I., Asuncion, A., Smyth, P., Welling, M.: Fast collapsed gibbs sampling for latent dirichlet allocation. In: KDD, pp. 569–577 (2008)
17. Shannon, C.E.: Prediction and entropy of printed english. The Bell System Technical Journal (1951)
18. Kumar, S., Barbier, G., Abbasi, M.A., Liu, H.: TweetTracker: An Analysis Tool for Humanitarian and Disaster Relief. In: ICWSM (2011)

Analog Textual Entailment and Spectral Clustering (ATESC) Based Summarization

Anand Gupta[1], Manpreet Kathuria[2], Arjun Singh[2], Ashish Sachdeva[2], and Shruti Bhati[1]

[1] Dept. of Information Technology, NSIT, New Delhi
omaranand@nsitonline.in, shrutibhati0@yahoo.co.in
[2] Dept. of Computer Engineering, NSIT, New Delhi
{Manpreet.kathuria,ashish.asachdeva}@gmail.com,
singh.arjun1313@nsitonline.in

Abstract. In the domain of single document and automatic extractive text summarization, a recent approach Logical TextTiling (LTT) has been proposed [1]. In-depth analysis has revealed that LTTs performance is limited by unfair entailment calculation, weak segmentation and assignment of equal importance to each segment produced. It seems that because of these drawbacks, the summary produced from experimentation on articles collected from New York Times website has been of poor/inferior quality. To overcome these limitations, the present paper proposes a novel technique called ATESC(Analog Textual Entailment and Spectral Clustering) which employs the use of analog entailment values in the range [0,1], segmentation using Normalized Spectral Clustering, and assignment of relative importance to the produced segments based on the scores of constituent sentences. At the end, a comparative study of results of LTT and ATESC is carried out. It shows that ATESC produces better quality of summaries in most of the cases tested experimentally.

Keywords: Textual Entailment, Spectral Clustering, Text Segmentation.

1 Introduction

According to [2], a summary can be defined as "A reductive transformation of source text to summary text through content reduction by selection and/or generalization on what is important in the source". Summarization based on content reduction by selection is termed as Extraction (identifying and including the important sentences in the final summary), whereas the one involving content reduction by generalization is termed as Abstraction (reproducing the most informative content in a novel way). The present paper focuses on Extraction based summarization. In this Section we discuss the earlier significant work in the field of automatic text summarization; identify their limitations and provide the means to overcome them.

S. Srinivasa and V. Bhatnagar (Eds.): BDA 2012, LNCS 7678, pp. 101–110, 2012.
© Springer-Verlag Berlin Heidelberg 2012

1.1 Related Work

In the field of Extraction based summarization, heuristics have been applied to identify the most important fragments of text (sentences, paragraphs etc.) in the context of the entire document. They are then selected to produce the summary. One of the most recent approaches to summarization that selects sentences as fragments and makes use of Logical TextTiling for segmentation has been proposed in [1]. We strongly feel that with certain modifications that our method proposes over the one using LTT, the quality of summaries produced for texts ranging across a variety of domains would considerably improve.

1.2 Limitations of LTT and the Proposed Improvements

1.2.1 Entailment Calculation

Textual Entailment is defined as the task of recognizing, given two pieces of text, whether meaning of one text can be inferred/entailed from the other. Recognizing Textual Entailment (RTE) has been proposed in 2005 as a generic task to build systems which capture the semantic variability of texts and performs natural language inferences [3]. In LTT [1] the textual entailment calculation is based on the sole criteria that "A text T entails a text H iff the text H is less informative than T". One sentence (piece of text) being less informative than the other is by no means a judgmental parameter in deciding whether the meaning of a sentence can be inferred from another. In LTT, the entailment between any pair of sentences is denoted by binary (0 or 1) values, where 0 means no entailment and 1 means no entailment.

We propose to replace this by analog entailment values lying between zero and one (both included), representing extent of entailment between each pair of sentences. This value is determined using VENSES [4] (Venice Semantic Evaluation System). The use of analog entailment values has strengthened the calculation of importance of the sentences.

1.2.2 Segmentation Criteria

In LTT a new segment is created whenever there is a change in the configuration of logical importance of sentences in the given text. The segmentation methodology employed is guilty of considering that a sentence is related in connotation to sentences only in its vicinity with respect to the text. The summary produced by LTT may be claimed to provide superior results in [1], but the very fact that the experiments are conducted on a single source of text is testimonial in proving that such a segmentation methodology cannot be generalized to a variety of documents/texts from different domains. We replace the LTT method of producing segments by the Normalized Spectral Clustering Algorithm suggested by Ng et. al. in [5]. The prime intuition of clustering [6], as the name suggests, is to separate a given number of points (elements) into distinct sets, such that:

1. The element segregated into any of the sets has maximum similarity amongst the elements of the same set.
2. The elements in different sets are most dissimilar to each other.

1.2.3   Segment Selection Criteria

LTT assigns equal importance to every segment, without giving any justification. It leads to the loss of important information as different segments show varying degree of relative important. In our approach, each segment is assigned a score equal to the mean score of the sentences comprising that segment. The segment score values are further used to rank the segments in order of merit.

1.3   Organization

The paper is organized as follows: Section 2 explains our proposed system architecture. Section 3 describes the preliminaries and the experimental results obtained, and Section 4 concludes the paper and enlists future work.

2   ATESC Architecture

The ATESC framework (Fig. 1) is broadly grouped into five phases as follows.

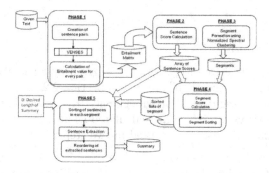

Fig. 1. System Architecture of ATESC

2.1   Analog Entailment Calculation

Input: The text document to be summarized.
Output: Entailment Matrix E[1:N,1:N], where N ñumber of sentences in the given text and an entry E[i,j] in the matrix denotes the extent by which the ith sentence en-tails the jth sentence in terms of Analog Entailment values.
Working Details: The given text is divided into pairs of sentences. The Analog Entailment value for every sentence pair is determined by employing the tool VENSES [4].

2.2   Sentence Scoring

Input: Entailment Matrix E[1:N,1:N] created in Section 2.1.
Output: An array of Sentence Scores i.e. SentenceScore[1:N], where SentenceScore[i] specifies the score of that sentence in the given text.

Working Details: Sentence score refers to the importance of a sentence and is equal to the sum of Analog Entailment values of all sentences which are entailed by the current sentence added to the sum of Analog Entailment values of all sentences which entail the current sentence. Thus a sentence with a higher score aptly represents a large number of sentences present in the given text/document. The mathematical expression for sentence score determination is given below.

$$\text{SentenceScore } [\, i \,] = \sum_{j=1}^{j=N} E\,[j,i] \; + \sum_{j=1}^{j=N} E\,[i,j] \qquad\qquad (1)$$

2.3   Text Segmentation/Clusterig

Input: Entailment Matrix E[1: N, 1: N] created in Section 2.1
Output:A set of segments Segment [1:K], where K is the number of segments created and depends on the text under consideration.
Working Details: The Entailment Matrix E[1:N, 1:N] represents the similarity matrix required as input to the Normalized Spectral Clustering algorithm given in [5]. The algorithm creates clusters (or segments) of sentences, which are most related to each other on the basis of their Analog Entailment values.

2.4   Segment Ranking

Input: Set of segments created in Section 2.3, given by Segment[1:K]; Array of sentence scores SentenceScore[1:N] created in Section 2.2.
Output: Sorted list of segments, SortedSegment[1:K] arranged in descending order of SegmentScore values which are calculated as mentioned below.
Working Details: The phase is executed in two steps.

1. Each segment is assigned a score equal to the mean score of all the sentences com-prising that segment as shown below:
   $$\text{SegmentScore}[i] = (1/L) \left\{ \sum_{j=1}^{j=L} SentenceScore[j] \right\} \qquad\qquad (2)$$
   where L is the number of sentences in segment i , where i lies in [1,K].
2. The segments are sorted in decreasing order of SegmentScore values calculated previously to produce the SortedSegment[1:K] array.

2.5   Sentence Extraction

Input: Array of sentence scores SentenceScore[1:N] created in Section 2.2; Sorted list of segments, SortedSegment[1:K] created in Section 2.4; Desired length of summary D. (D is the number of sentences in the final summary).
Output: Summary generated by ATESC based Summarization of desired length D.
Working Details: The phase is executed in three steps.

1. Sentence Sorting within every segment: The sentences in each of the segments are arranged in the decreasing order of SentenceScore[1:L] values (L is the number of sentences in a segment).

2. Sentence Selection: The sentence with highest SentenceScore value is selected from each of the segments beginning from the first segment in the SortedSegment[1:K] array, until the number of extracted sentences equals the desired length of summary to be produced. If the desired length of summary exceeds the total number of segments created by ATESC, i.e. D is greater than equal to K then the sentence with the second highest value of SentenceScore is selected from each of the segments, again beginning from the first segment in the SortedSegment[1:K] array. In this manner we keep on selecting sentences for our summary from the sentences of the next highest value of SentenceScore from various segments in the SortedSegment[1:K] array until the length of the produced summary equals the desired length D.
3. Sentence Reordering: All the sentences selected are arranged according to their order in the original document and the final summary is produced.

3 Experimental Results and Discussions

3.1 Datasets

Experiments are conducted on a collection of 40 text documents comprising of news articles and classic essays collected from the internet [7]. The lengths of the articles vary from 30 to 105 sentences.

3.2 Baselines and Evaluation Parameters

Performance evaluation of a summarization tool is a very difficult task since there is, in fact, no ideal summary. It has been proved in [8] that even for simple news articles, human made summaries tend to agree with the automated summaries only 60% of the time. Therefore, we consider two baselines, Microsoft Word 2007 AutoSummarize [9] and Human Developed summaries to compare summaries produced by ATESC based summarization against those produced by LTT based summarization. The summaries produced by both the above methods (ATESC and LTT) are called candidate summaries while those produced using the mentioned baselines are called reference summaries. The following three parameters have been used to analyze the quality of the summaries produced Precision, Recall and F-measure denoted by P, R and F respectively and expressed in percentage (%). P, R and F values are computed in the context of text summarization as shown in the following formulae:

$$P = \frac{|\{\#\text{sentencescand}\} \cap \{\#\text{sentencesref}\}|}{|\{\#\text{sentencescand}\}|} \qquad (3)$$

$$R = \frac{|\{\#\text{sentencescand}\} \cap \{\#\text{sentencesref}\}|}{|\{\#\text{sentencesref}\}|} \qquad (4)$$

$$\text{and} \quad F = \text{Harmonic mean of P and R.} \qquad (5)$$

"#" means number of; "cand" means candidate; "ref" means reference.

3.3   Performance Evaluation and Discussion

There are three major differences in the summarization methods using ATESC and LTT which are given below.

1. Calculation of Entailment Matrix Analog (ATESC) vs. Binary (LTT)
2. Segmentation Spectral Clustering(ATESC) vs. Logical TextTiling(LTT)
3. Segment Scoring Mean Sentence Score (ATESC) vs. Assumed Equal (LTT)

On the basis of above there can be a total of 8 (2*2*2) combinations 2 for each of those phases where ATESC is distinct from LTT. These eight combinations are denoted as numerals from 1 to 8, and are described in detail below.
  Each of the above combinations is experimented with both the baselines in order

Table 1. All combinations of methods. Analog = [0,1]. LTT:Logical TextTiling[1]. NSC:Normalized Spectral Clustering [5]. AE:Assumed Equal. MS: Mean Sentence Score. The combinations used in the graphs are as per the method codes given in this table (Column 1).

Method Code	Entailment Matrix	Segmentation	Salience
1	Binary(0/1)	LTT	AE
2	Binary(0/1)	LTT	MSS
3	Binary(0/1)	NSC	AE
4	Binary(0/1)	NSC	MSS
5	Analog	LTT	AE
6	Analog	LTT	MSS
7	Analog	NSC	AE
8	Analog	NSC	MSS

to observe the effect, which each of the three changes proposed in this paper has on the quality of the final summary produced. We have considered four distinct cases to conduct the experiments, and their results are discussed as follows: (The numbers in curly braces, as they appear in the following text are used to depict a bar graph in one of the figures, which is denoted by the same number. For e.g. ({4} in Fig. 13) refers to the bar graph of the method having Method Code 4 as per Table 1 in Figure 13).

3.3.1   Different Length of Input Text
The text documents in our dataset are classified into two categories according to their lengths Short and Long.

Short Texts. The results for this category are shown in Figs. 2 and 3. With MS Word AutoSummarize as the baseline, it is observed that the most accurate summaries ({6}in Fig. 2) are generated when the segmentation by LTT method is based on Analog Entailment values and is succeeded by the assignment of segment scores and not assuming them to be equal for all segments. Both of these changes are employed in ATESC based summarization.

LTT revolves around the notion that sentences are most closely related when at shorter distances to each other. ATESC is independent of such a presumption. But for shorter texts the distances among sentences is not very large. Thus the characteristic feature of ATESC does not get an opportunity to be tested, pertaining to shorter distances among sentences in a Short Text. As a result for Short Texts LTT is observed to produce more accurate summaries.

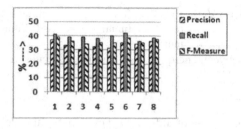

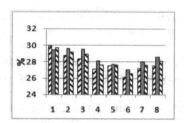

Fig. 2. P, R, F for Short Texts MS Word 2007 Auto Summarize

Fig. 3. P, R, F for Short Texts Human Developed Summaries

Long Texts. The results for this category are shown in Figs. 4 and 5. The notion of distant sentences exhibiting strong relationship with each other (with respect to connotation) is proved by the results of experimentation on Long Texts. From Figs. 4 and 5, it is observed that the replacement of LTT methodology by ATESC methodology (transmute from {1} to {8} in Figures 4 and 5), we observe a gradual and proportional rise in the quality of summaries produced. The quality increases almost linearly for both the baselines employed but with minor differences. Also in both the graphs the best summary is generated by employing ATESC in entirety ({8} in both Figures 4 and 5).

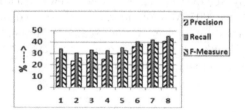

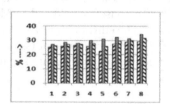

Fig. 4. P, R, F for long Texts MS Word 2007 Auto Summarizer

Fig. 5. P, R, F for long Texts Human Developed Summaries

3.3.2 Jumbling the Sentences of a Given Text

Jumbled text refers to the text document in which the original order of sentences is changed randomly. The objective of jumbling up a given text and then summarizing it is to address those text documents where contextual sentences (sentences closely related to each other) are distributed erratically within the text. A prime example of such texts is e-mails. It is observed from Figs. 6 and 7

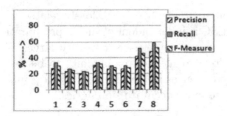
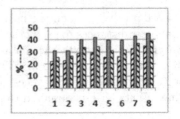

Fig. 6. P, R, F for Jumbled texts MS Word 2007 Auto Summarizer

Fig. 7. P, R, F measure for Jumbled texts - Human Developed Summaries

that there is a prominent difference between the quality of summaries generated by ATESC and LTT in favor of ATESC.

3.3.3 Input Text Classification

The dataset is intentionally made to comprise of different types of texts in order to study the performance of all the methods in detail. We classify our input texts in two categories: Essays and News Articles.

Essays. The articles belonging to this category follow a logical structural organization. For such kind of articles, LTT is bound to give better results compared to ATESC, which is proved experimentally and demonstrated by graphs in Figs. 8 and 9. It is also inferred that with MS Word 2007 AutoSummarize as reference (Fig. 8), the results are most accurate when LTT method is integrated with analog entailment values rather than discrete ({5} in Fig. 8) or when the segment scoring is employed after segmentation using LTT ({2} in Fig. 8); both of these changes are proposed in ATESC. Furthermore, when we consider Human generated summary as a base, there is a very thin line of difference between the results of ATESC and LTT as shown in Fig. 9.

News Articles. ATESC performs fairly well over LTT in the case of news articles when the baseline selected is MS Word 2007 AutoSummarize as illustrated in Fig. 10. The editor of a news article generally connects various sentences, present in different paragraphs of the article. Unlike LTT, ATESC can easily determine such affiliations and create segments comprising of such sentences.

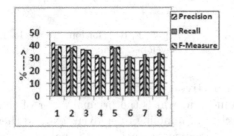
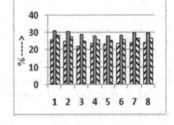

Fig. 8. P, R, F for Essays - MS Word 2007 Auto Summarizer

Fig. 9. P, R, F for Essays - Human Developed Summaries

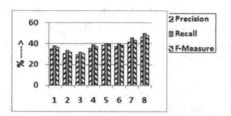

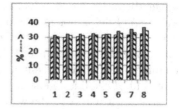

Fig. 10. P, R, F for News Articles MS Word 2007 Auto Summarizer

Fig. 11. P, R, F for News Articles - Human Developed Summaries

3.3.4 Different Summary Lengths of the Same Text

Summaries measuring 25%, 35% and 45% in length of the entire text are produced using ATESC and LTT with the MS Word 2007 AutoSummarize as reference. It can be observed from Figs. 12 and 13 that on increasing the summary length for a given text, both ATESC and LTT based summarizations exhibit improvements, though these improvements are more significant in ATESC. This is ascertained to the fact that the introduction of assignment of scores to the segments created empowers ATESC to select more appropriate sentences as the length of summary is increased. In case of LTT, since each segment is assigned equal importance, on increasing the length of the summary, the inclusion of the next most appropriate sentence is unlikely.

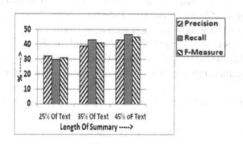

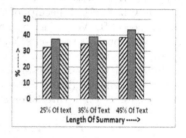

Fig. 12. P, R, F for different summary rates using ATESC

Fig. 13. P, R, F for different summary rates using LTT

4   Conclusions and Future Work

In this paper, we have presented a new approach to Text Summarization based on Analog Text Entailment and Segmentation using Normalized Spectral Clustering. It has shown best results for texts which are long and belong to the domain of texts/documents where relationship among sentences is independent of their respective positions, such as News Articles and e-mails. However, in texts like essays,which are more structured, the best results are observed when the LTT segmentation methodology is supplemented with one of the changes proposed by ATESC. In case of texts/documents, where the likelihood of sentences at larger distances being related is constrained by the length of the text (Short Texts) ATESC is not found to be the most effective Summarization approach. The increase in the length of the produced summaries has a greater influence in improving the performance in case of ATESC than in LTT pertaining to the introduction of mean segment score values in the former. It can thus be concluded that ATESC is able to effectively overcome the short-comings of LTT. It is safe to say that it will pave the way for future research and experimentation towards the use of analog entailment values and clustering algorithms for segmentation of text.

It may be mentioned that the performance of ATESC based Summarization is limited by the effectiveness and computation speed of the employed Textual Entailment Engine and the Segmentation Algorithm. An improvement in either of these is sure to enhance the quality of the summary produced.

References

1. Tatar, D., Tamaianu-Morita, E., Mihis, A., Lupsa, D.: Summarization by Logic Segmentation and text Entailment. In: The Proceedings of Conference on Intelligent Text Processing and Computational Linguistics (CICLing 2008), Haifa, Israel, February 17-23, pp. 15–26 (2008)
2. Jones, K.S.: Automatic summarizing: The state of the art. Information Processing & Management 43(6), 1449–1481 (2007)
3. Iftene, A.: Thesis on AI, Textual Entailment, TR 09-02, University "Alexandru Ioan Cuza" of Iasi, Faculty of Computer Science (October 2009)
4. Delmonte, R.: Venses, http://project.cgm.unive.it/venses_en.html
5. Ng, A., Jordan, M.I., Weiss, Y.: On Spectral Clustering: Analysis and an algorithm. In: Dietterich, T., Becker, S., Ghahramani, Z. (eds.) The Advances in Neural Information Processing Systems, Vancouver, British Columbia, Canada, December 3-8, pp. 849–856 (2001)
6. Luxburg, U.V.: A Tutorial on Spectral Clustering. Journal Statistics and Computing 17(4), 1–32 (2007)
7. Classic Essays, http://grammar.about.com/od/classicessays/ CLASSIC_ESSAYS.html, News Articles, http://www.nytimes.com/
8. Radev, D.R., Hovy, E., McKeown, K.: Introduction to the Special issue on Summarization. Journal Computational Linguistics 28(4), 399–408 (2002)
9. Microsoft Auto Summarizer 2007 is used to identify the key points in the document and create a summary, http://office.microsoft.com/en-us/word-help/ automatically-summarize-a-document-HA010255206.aspx

Economics of Gold Price Movement-Forecasting Analysis Using Macro-economic, Investor Fear and Investor Behavior Features

Jatin Kumar[1], Tushar Rao[2], and Saket Srivastava[1]

[1] IIIT-Delhi
[2] Netaji Subhas Institute of Technology- Delhi
{saket,jatin09021}@iiitd.ac.in, rao.tushar@nsitonline.in

Abstract. Recent works have shown that besides fundamental factors like interest rate, inflation index and foreign exchange rates, behavioral factors like consumer sentiments and global economic stability play an important role in driving gold prices at shorter time resolutions. In this work we have done comprehensive modeling of price movements of gold, using three feature sets, namely- macroeconomic factors (using CPI index and foreign exchange rates), investor fear features (using US Economy Stress Index and gold ETF Volatility Index) and investor behavior features (using the sentiment of Twitter feeds and web search volume index from Google Search Volume Index). Our results bring insights like high correlation (upto 0.92 for CPI) between various features, which is a significant improvement over earlier works. Using Grangers causality analysis, we have validated that the movement in gold price is greatly affected in the short term by some features, consistently over a five week lag. Finally, we implemented forecasting techniques like expert model mining system (EMMS) and binary SVM classifier to demonstrate forecasting performance using different features.

Keywords: Gold, Twitter, SVI, Stress Index, VIX, Forex rates.

1 Introduction

Most of the academic or industrial work on gold prices make use of only fundamental macro-economic factors without taking into account the investor perceptions and sentiments towards gold and economy in general. We analyze the gold prices using three types of factors- macro-economic factors, factors depicting level of economic stability or investor fear features and the investor behavior features that we get from Twitter and Search Volume Index (Google SVI). We apply techniques like pearson correlation, cross-correlation at different time lags and Grangers causality analysis for establishing a causative relationship between the gold prices and the above mentioned features. Further, we use forecasting techniques like EMMS and SVM to predict gold prices and the direction of its movement. To the best of our knowledge, there is no earlier work that deals

S. Srinivasa and V. Bhatnagar (Eds.): BDA 2012, LNCS 7678, pp. 111–121, 2012.

with the comprehensive comparison between the investor perception and sentiment with macro-economic factors [5, 8–10]. This paper marks a novel step in the direction of analyzing the effect of gold prices on macro-economic factors in combination with investor's perception or behavior. Using the macro-economic factors, we try to establish a long-term pattern of these factors with the gold prices, while we use Twitter feed data, Google SVI or gold ETF volatility index to study the short-term fluctuations in gold prices.

Figure 1 summarizes the work flow of the paper. In section 2 we present data collection and prior processing, defining the terminologies used in the market securities. Further, in section 3 we present the statistical techniques implemented and subsequently discuss the derived results.

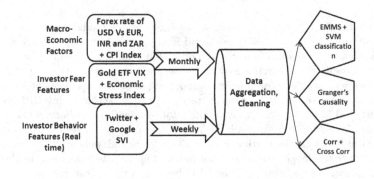

Fig. 1. Work flow of the techniques implemented in this research work

2   Dataset Preparation and Mining

We have made use of spot gold price data from World gold Council at two levels of analysis- monthly for macroeconomic and investor fear features and weekly for investor behavior features [1]. Further, in this section we discuss the collection and processing of the time series data used in this work.

2.1   Macro-economic Factors

We use monthly time series for the macroeconomic factors - Consumer Price Index (CPI) and forex rates of USD with major currencies. Time period of analysis in this features set is from January 1st 2003 to January 1st 2012. This time period corresponds to one of the robust changes in the economy, including meteoric rise in 2004 and recession in 2008 due to sub-prime mortgage crisis. This provides us more confidence in the causative relations and the conclusions drawn from the statistical results. The CPI in the United States is defined by the Bureau of Labor Statistics as "*a measure of the average change over time*

---

[1] http://www.gold.org/investment/statistics/goldpricechart/

in the prices paid by urban consumers for a market basket of consumer goods and services." Due to constrictive feature space, we have taken forex rates of dollar versus major traded (Euro) and major consumer and producer countries i.e. India (INR) and South Africa (ZAR) respectively. Exchange rates of USD vs EURO, INR and ZAR are collected from Federal Reserve Bank of St. Louis[2].

2.2 Economic Stability and Investor Fear Features

Primarily due to two reasons, at a smaller resolution say weekly, macro-economic factors are not suffice to provide clear explanation for noisy patterns in the price. Firstly, these factors are not available at this small frequency, for example CPI index is calculated only at monthly frequency. Secondly, macro-economic factors doesn't reveal anything about the investor and economy perception of the near future. We have made use of two time series for this segment- gold ETF Volatility Index (VIX) and the St. Louis Financial Stress Index (measures economic stability). Time period of analysis in this features set is from June 16th 2008 to February 3rd 2012 (each Friday). CBOE gold VIX is a index measure of the market expectation of 30-day ahead volatility of gold price.

2.3 Realtime Market Sentiment or Investor Behavior Features

In this section, we discuss the extraction of data from Twitter and Google Insights for Search. This includes, collecting search volume indices for various search queries related to gold using Google Insights for Search and mining of tweets- processing them for analysis, extracting tweet sentiment and normalizing them for forecasting. Time period of analysis in this features set is weekly, from June 2nd 2010 to September 13th 2011 (each Friday).

Tweets. Tweets are made aggregated through a simple search of keywords for gold (like $GLD) through an application programming interface (API)[3]. In this paper, we have used tweets from period of 15 months and 10 days between June 2nd to 13th September 2011. During this period, by querying the Twitter search API for gold, we collected $1,964,044$ (by around 0.71M users) English language tweets. In order to compute sentiment for any tweet, we classify each incoming tweet into *positive* or *negative* using Naive Bayesian classifier. For each week, total number of positive tweets is aggregated as $Positive_{week}$ while the total number of negative tweets as $Negative_{week}$. We have made use of JSON API from Twittersentiment[4], a service provided by Stanford NLP research group [4]. Online classifier has made use of Naive Bayesian classification method, which is one of the successful and highly researched algorithms for classification giving superior performance to other methods, in context of tweets.

---

[2] Federal Reserve Economic Data: http://research.stlouisfed.org/fred2/

[3] Twitter API is accessible at https://dev.twitter.com/docs

[4] https://sites.google.com/site/twittersentimenthelp/

Feature Extraction and Aggregation. We have selected weekly domain as time resolution under study over daily, bi-daily, bi- weekly or monthly as it is the most balanced window to study effect of investor behavior and model performance accuracy, keeping in-market monetization potential practically impeccable. For every week, the value of closing price of gold is recorded, every Friday. To explore the relationships between weekly trading and on days when market remains closed (weekends, national holidays) we broadly focus on two domains of tweet sentiments-weekday and weekend. We have carried forward work of Antweiler et al. [2] for defining bullishness (B_t) for each time domain given by equation:

$$B_t = \ln \frac{1 + M_t^{Positive}}{1 + M_t^{Negative}} \qquad (1)$$

Where $M_t^{Positive}$ and $M_t^{Negative}$ represent number of positive or negative tweets during particular time period. Logarithm of bullishness measures the share of surplus positive signals and also gives more weight to larger number of messages in a specific sentiment (positive or negative). Message volume is simply defined as natural logarithm of total number of tweets per time domain for a specific security/index. And the agreement among positive and negative tweet messages is defined as:

$$A_t = 1 - \sqrt{(1 - \frac{(M_t^{Positive} - M_t^{Negative})}{(M_t^{Positive} + M_t^{Negative})}} \qquad (2)$$

If *all* tweet messages about gold prices are bullish or bearish, agreement would be 1 in that case. Influence of silent tweets days in our study (trading days when no tweeting happens about particular company) is less than 0.1%, which is significantly less than previous research [2,7]. Thus, for the gold asset we have a total of three potentially causative time series from Twitter the bullishness, message volume and agreement.

Search Volume Index. To generate search engine lexicon for gold, we tested by collecting weekly search volume for specific search terms related to gold such as 'buy gold', 'invest in gold' etc. from **Google Insights of Search**[5]. Further, have ease in the computation, by applying dimension reduction technique of principle component analysis; we are able to reduce the number of variables (uptil 8 for gold) from search domain by combining similarly behaving time series creating completely uncorrelated co-independent factors- Fact 1 and Fact 2 in our case.

3   Statistical Techniques and Results

First we identify correlation patterns between various time series at different lagged intervals, further testing the causative relationships of various features with the gold price using econometric technique of Grangers Casuality Analysis (GCA). Then we make use of forecasting techniques like expert model mining system (EMMS) and binary- SVM classifier to propose and test the forecasting model and draw performance related conclusions.

---

[5] http://www.google.com/insights/search/

3.1   Correlation and Cross-Correlation Analysis

We begin our study by identifying pairwise pearson correlation between the respective feature sets in three categories- macroeconomic, investor fear and investor behavior with the gold prices. As an evaluation of lagged response of relationships existing between macro-economic factors, Twitter sentiments and search volumes, we compute cross-correlation at lag of $pm7$ week lag to show effectiveness in prediction and motivate us to look forward in making an accurate forecasting model by picking accurate regressor co-efficient. For any two series $x = \{x_1, ..., x_n\}$ and $y = \{y_1, ..., y_n\}$, the cross correlation lag γ at lag k is defined as:

$$\gamma = \frac{\sum_i (x_{i+k} - \overline{x})(y_i - \overline{y})}{\sqrt{\sum_i (x_{i+k} - \overline{x})^2} \sqrt{\sum_i (y_i - \overline{y})^2}} \qquad (3)$$

In equation 3, $\overline{x}$ $\overline{y}$ are the mean sample values of x and y respectively. Cross-correlation function defined as short for ccf(x,y), is estimate of linear correlation between x_{t+k} and y_t, which means keeping the time series y stationary, we move the time series y backward to forward in time by a lag of k i.e. k= [-7,7] for lags for 7 weeks in positive and negative direction. The Tables 1 contains the summarized set of results for three feature time series after transformation to log scale. Amongst macroeconomic factors we observe, highest correlation i.e. 0.924 of gold price is with CPI. This indicates surge in the price index of gold when CPI increases. While amongst three different forex rates, correlation value of -0.589 is observed with USD to EUR while its positive with ZAR and INR. This indicates, gold price is not governed by the buyer or the supplier end, but due to extensive market conditions in the European region. While amongst investor fear features, financial stress index bears a very high value of 0.618 with the gold price. This gives us plausible explanation of - why people rush to invest in gold in times of unstable market conditions, causing the surge in the price. While amongst investor behavior features, only agreement and search volume index bear high degree of correlation. We can observe that, when there is less agreement between the positive and negative time series on Twitter, price of gold rises again as it is seen as the most safe investment in unsure market conditions. There is heavy increase in the Google searches when the price is going up, probably because people are talking more about it.

Further as we can see from the figure 2, correlation values over various positive and negative lags depict varied relationship. There is high positive (or negative) correlation between gold and CPI, gold and stress index and gold and Google SVI at different lags. Correlation values taken at any lags does not always necessarily mean a causative relationship. We perform a more robust test for inferring a causative relation by performing Grangers Causality Analysis (GCA), as discussed in next section.

Table 1. Correlation between gold price (G) vs Macroeconomic Features, Investor Behavior Features and Investor Sentiment Features

Features	CPI	USD/ ZAR	USD/ INR	USD/ EUR	Stress Index	ETF VIX	Bull	Msg Vol.	Agrmnt	SVI
gold	.924	.309	.278	-.589	-.618	-.446	-.352	.137	-.586	.726

3.2   Grangers Causality Analysis

GCA rests on the assumption that if a variable X causes Y then changes in X will be systematically occur before the changes in Y. We realize lagged values of X shall bear significant correlation with Y. In line with the earlier approaches by [1, 3] we have made use of GCA to investigate whether one time series is significant in predicting another time series. GCA is used not to establish statistical causality, but as an economist tool to investigate a statistical pattern of lagged correlation. A similar observation is that, smoking causes lung cancer is widely accepted; proving it contains carcinogens but itself may not be actual causative of the real event.

Let returns R_t be reflective of fast movements in the stock market. To verify the change in returns with the change in Twitter features we compare the variance given by following linear models in equation 4 and equation 5.

$$R_t = \alpha + \Sigma^n_{i=1}\beta_i D_{t-i} + \epsilon_t \tag{4}$$

$$R_t = \alpha + \Sigma^n_{i=1}\beta_i D_{t-i} + \Sigma^n_{i=1}\gamma_i X_{i-t} + \varepsilon_t \tag{5}$$

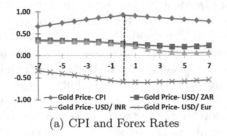

(a) CPI and Forex Rates

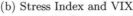
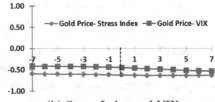

(b) Stress Index and VIX

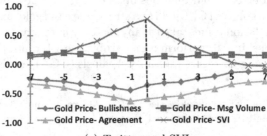

(c) Twitter and SVI

Fig. 2. Cross Correlation of gold price vs various causative features like CPI & Forex rates (a) and Stress Index & VIX (b); and Twitter & SVI (c)

Table 2. Granger's Casuality Analysis- statistical significance (p values) at lags of 0-5 weeks between gold Price and Macroeconomic Features

Lag Week	gold Price-CPI	gold Price-USD/ZAR	gold Price-USD/INR	gold Price-USD/EUR
Lag 0	<.001	.004	.008	<.001
Lag 1	<.001	.007	.027	<.001
Lag 2	<.001	.012	.087	<.001
Lag 3	<.001	.019	.196	<.001
Lag 4	<.001	.029	.337	<.001
Lag 5	<.001	.030	.450	<.001

Table 3. Granger's Casuality Analysis- statistical significance (p values) at lags of 0-5 weeks between gold Price and Investor Behavior Features

Lag Week	gold- Stress Index	gold- VIX
Lag0	<.001	<.001
Lag1	<.001	<.001
Lag2	<.001	<.001
Lag3	<.001	<.001
Lag4	<.001	<.001
Lag5	<.001	<.001

Table 4. Granger's Casuality Analysis- statistical significance (p values) at lags of 0-5 weeks between gold Price and Investor Sentiment Features

Lag Week	gold Price-Bullishness	gold Price-Msg Volume	gold Price-Agreement	gold Price-SVI
Lag 0	.113	.065	.002	<.001
Lag 1	.167	.068	.003	<.001
Lag 2	.248	.053	.007	<.001
Lag 3	.522	.023	.032	<.001
Lag 4	.695	.019	.068	<.001
Lag 5	.890	.012	.213	.001

Equation 4 uses only 'n' lagged values of R_t, i.e. $(R_{t-1}, \ldots, R_{t-n})$ for prediction, while equation 5 uses the n lagged values of both R_t and the various attribute of the time series given by $X_{t-1}, \ldots, X_{t-n}$.

If p-values are significant (say less than 0.05) and show an increasing trend as we go down the time for some time lags, causative relation between the time series can be inferred. Observing the Grangers causality results in the tables above, there is no significant causal relationship of gold with USD/INR forex, which could answered by the fact that although India is the largest consumer of gold, but most of the gold transactions still happen in US dollars. On the other hand, USD/ZAR do have a causal relationship to gold prices considering the fact that processes that are employed in extraction of gold from mines happen in local currency and hence gold price would change as price of local currency varies.

Gold price also show a causal relationship with the agreement feature, that represents the vector distance between positive and negative sentiments about the gold prices. This is intuitive, since even if all the tweets about gold are negative (eg gold on a all-time low), then people tend to buy gold and gradually causing the price of gold to rise. Similarly, if all the tweets are positive, even then demand gets increased and prices tend to rise. In the case of Google SVI, there is a significant causal relation with the gold prices, which is also intuitive since searches like "buy in gold", "invest in gold" are used to compute the search volume index.

3.3   EMMS Model for Forecasting

Selection criterion for the EMMS is MAPE and stationary 'R squared', which is the measure of how good is the model under consideration on comparison to the baseline model [6]. To show that in prediction model, we have applied the EMMS twice - first with the various features as independent predictor events and second time, without them. This provides us with a quantitative comparison of improvement in the prediction when we use the features under consideration. ARIMA (p,d,q) are in theory and practice, the most general class of models for forecasting a time series data, which is subsequently stationarized by series of transformation such as differencing or logging of the series Y_i. In the dataset, we have time series for a total of approximately 9 years (108 months) for macroeconomic factors, 188 weeks for economic stability factors and 60 weeks for investor behavior factors, out of which we use approximately 75% of the time period for the training both the models with and without the predictors. Further, we verify the model performance as one step ahead forecast over the testing period of 27 months for macroeconomic factors, 47 weeks for economic stability factors and 15 weeks for investor behavior factors. The testing time period account for wide and robust range of market conditions. Model equation for two cases are given below as equation 6 for forecasting without predictors and equation 7 for forecasting with predictors. In these equations Y represents gold price and X represents the investor behavior series from SVI and Twitter features.

$$WithoutPredictors: Y_t = \alpha + \Sigma^n_{i=1} \beta_i Y_{t=i} + \epsilon_t \qquad (6)$$

$$WithPredictors: Y_t = \alpha + \Sigma^n_{i=1} \beta_i Y_{t=i} + \Sigma^n_{i=1} \gamma_i X_{t=i} + \epsilon_t \qquad (7)$$

Table 5. Average MAE percentage using EMMS

Model Type	Predictors	
	Yes	No
Macroeconomic Features	4.00	4.04
Investor Behavior Features	2.16	2.32
Investor Sentiment Features	1.52	1.59

It is observed that the prediction accuracy improves in all the cases when we include the predictors in the EMMS model. The most significant improvement is in the case when we take into account gold ETF VIX and Financial Stress Index as parameters in the forecasting equation of the gold price.

3.4   SVM

We ran simulations to check out the effectiveness of the proposed three features in predicting direction of the gold price movement for the next day. Approximately, 75% of the time period is taken in the training phase to tune the SVM classifier (using features from the prior week). The trained SVM classifier is then used to predict market direction (price index movement) in the coming week. Testing

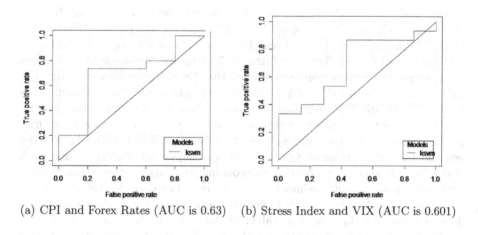

(a) CPI and Forex Rates (AUC is 0.63)   (b) Stress Index and VIX (AUC is 0.601)

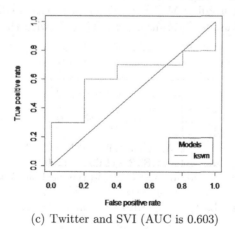

(c) Twitter and SVI (AUC is 0.603)

Fig. 3. ROC curves for directional accuracy of gold Price vs various causative features like CPI & Forex rates (a) and Stress Index & VIX (b); and Twitter & SVI (c)

phase for the classification model (class 1- bullish market ↑ and class 0- bearish market ↓) is 27 months for macroeconomic factors, 47 weeks for economic stability factors and 15 weeks for investor behavior factors. SVM model is build using kSVM classification technique with the *rbf* kernel using the package 'e1071' in R statistical language. Receiver operator characteristics (ROC) curve measuring the accuracy of the classifier as true positive rate to false positive rate is given in the figure 3. It shows the tradeoff between sensitivity i.e. true positive rate and specificity i.e. true negative rate (any increase in sensitivity will be accompanied by a decrease in specificity). Good statistical significance for the classification accuracy can be inferred from the value of area under the ROC curve (AUC) which comes out to 0.6.

4   Conclusion

In this work, we have analyzed the effect of gold prices using three different classes of factors- macro-economic factors (CPI index, USD/EUR, USD/INR and USD/ZAR forex rates), investor fear features (Economy Stress Index and gold ETF VIX) and investor behavior features (Twitter feeds, Google SVI). Evident from the results of the techniques applied that the macro-economic factors like CPI index, USD value, USD/ZAR forex play a significant role in determining the future of gold prices. This is mainly because these factors change the demand and supply forces operating in case of gold commodity. For example, change in CPI index causes the interest rates to change, which in turn changes the demand for gold and hence effects its prices. Important finding of this work is the role of investors perception of economic stability on the prices of gold. Although these are not fundamental factors but they do cause the gold prices to fluctuate significantly. This is mainly because gold is a speculative asset that does not have any fundamental value, unlike other assets such as companies' stocks or properties or other commodities used for industrial use like steel. This work shows a significant fall in MAPE for different robust prediction techniques for estimating direction and magnitude of gold price with the use of predictor variables.

References

1. Huina, M., Xiao, Z., Bollen, J.: Twitter mood predicts the stock market. Computer 1010(3003v1), 1–8 (2010)
2. Antweiler, W., Frank, M.Z.: Is All That Talk Just Noise? The Information Content of Internet Stock Message Boards. SSRN eLibrary (2001)
3. Karrie, K., Gilbert, E.: Widespread worry and the stock market. Artificial Intelligence, 58–65 (2010)
4. Bhayani, R., Huang, L., Go, A.: Twitter Sentiment Classification using Distant Supervision
5. Yagil, J., Qadan, M.: Fear sentiments and gold price: testing causality in-mean and in-variance. Applied Economics Letters 19(4), 363–366 (2012)

6. Laura, S.: Forecasting structural time series models and the kalman filter. Journal of Applied Econometrics 6(3), 329–331 (1991)
7. Welpe, I.M., Sprenger, T.O.: Tweets and Trades: The Information Content of Stock Microblogs. SSRN eLibrary (2010)
8. He, J., Yang, N.: Statistical investigation into the accelerating cyclical dynamics among gold, dollar and u.s. interest rate. In: Modeling Risk Management for Resources and Environment in China, pp. 67–76 (2011)
9. Fuehres, H., Gloor, P.A., Zhang, X.: Predicting stock market indicators through twitter i hope it is not as bad as i fear. Procedia - Social and Behavioral Sciences 26(0), 55–62 (2011)
10. Fuehres, H., Gloor, P.A., Zhang, X.: Predicting asset value through twitter buzz 113, 23–34 (2012)

An Efficient Method of Building an Ensemble of Classifiers in Streaming Data

Joung Woo Ryu[1], Mehmed M. Kantardzic[2], Myung-Won Kim[3], and A. Ra Khil[3]

[1] Technical Research Center, Safetia Inc., Seoul, 137-895, South Korea
[2] CECS Department, Speed School of Engineering, University of Louisville, KY 40292, USA
[3] Department of Computer Science, Soongsil University, Seoul, 156-743, South Korea
ryu0914@safetia.com, mmkant01@louisville.edu,
{mkim,ara}@ssu.ac.kr

Abstract. To efficiently refine a classifier in streaming data such as sensor data and web log data we have to decide whether each streaming unlabeled datum is selected or not. The exiting methods refine a classifier based on a regular time interval. They refine a classifier even if the classification accuracy of the classifier is high. Also it uses a classifier even if the classification accuracy is low. In this paper, our ensemble method selects data in an online process that should be labeled. The selected data are used to build new classifiers of an ensemble. Our selection methodology uses training data that are applied to generate an ensemble of classifiers over streaming data. We compared the results of our ensemble approach and of a conventional ensemble approach where new classifiers for an ensemble are periodically generated. In experiments with ten benchmark data sets including three real streaming data sets, our ensemble approach generated 12.9% new classifiers for the chunk-based ensemble approach using partially labeled samples, and used an average of 10% labeled samples for the ten data sets. In all the experiments, our ensemble approach produced comparable classification accuracy. We showed that our approach can efficiently maintain the performance of an ensemble over streaming data.

Keywords: Classification, Ensemble, Streaming data, Active learning.

1 Introduction

In a stream classification problem, the current classifier assigns a label out of a predefined set to each input sample. After classification, only when the correct labels of all new samples are available, the current classifier's performance can be evaluated and the current classifier can be refined. It is very impractical for a human expert to give the classifier feedback on its decision for every single sample. In a real-world application, a typical approach is to divide a data stream into subsets of streaming samples for periodically adjusting and improving a classifier using a fixed time interval. The subset of streaming samples is referred as a "chunk" or "batch." In applications detecting "cybercrime", classifiers decide in real time whether a behavior

S. Srinivasa and V. Bhatnagar (Eds.): BDA 2012, LNCS 7678, pp. 122–133, 2012.
© Springer-Verlag Berlin Heidelberg 2012

pattern is a crime pattern or not. The classifiers are usually refined periodically using a "chunk" of consecutive sequence samples. For example, let us suppose that one million samples of click streaming data are gathered every day, and a click fraud classifier is updated at night every 24 hours. If new types of click fraud patterns occur in streaming data, the current classifier may miss those new patterns until it is changed to accommodate them. In the above example, human experts must analyze one million samples of click streaming data every day to find new click fraud patterns. This process is very expensive and mostly impractical for the human experts. We believe that new click patterns do not occur every day. In order to maintain the performance of the current classifier in non-stationary environments, we should carefully consider the following two matters: (1) when human experts evaluate or refine the current classifier, and (2) which samples should be labeled for evaluating or refining the current classifier.

Online learning approaches can refine the current classifier whenever the correct labels of streaming data are obtained. Such an approach is able to quickly detect points in time where changes in streaming data distribution happen. Many researchers show such an advantage in their works [1,2]. They implicitly assume that all streaming data have true labels and that those correct labels can be used at any time. This assumption requires additional methods for efficiently applying online approaches to real-world applications. On the other hand, the conventional ensemble approach fundamentally stores streaming samples in a buffer. Whenever the buffer is full, all samples within the buffer are manually labeled, and a new classifier for an ensemble is built from those labeled samples. The ensemble approach has shown higher classification accuracy for streaming data with changes in data distribution than a single classifier approach [3,4].

The conventional ensemble approach works on the assumption that all streaming data have correct labels. The conventional ensemble approach assigns a new weight to each classifier of an ensemble or builds new classifiers for an ensemble using cross-validation whenever a new chunk is coming. Chu et al.[5] and Zhang et al.[6] used the weighted samples when a new classifier for an ensemble are built from a chunk. Zhang et al. [7] proposed an aggregate ensemble framework where different classifiers are built by each different learning algorithm from a chunk including noisy data. Wei et al.[8] proposed an ensemble approach for multi-label classification problems in streaming data where each sample can be classified into more than one category (e.g. multiple illnesses and interesting topics at the same time). However, it is not practical to have human generated labels of all streaming data in the real-world applications.

Recently, some researchers have recognized that it is not reasonable in practice to manually label all samples within a chunk for building a new classifier of an ensemble [9,10,11]. In particular, the ensemble approaches of Masud et al. [9] and Woolam et al. [10] randomly select a small amount of unlabeled samples from a chunk and use their correct labels. They proposed a "semi-supervised clustering" algorithm to create K clusters from a chunk where samples are partially labeled, and use a set of K clusters as a classification model. Masud et al. [9] find the Q nearest clusters from an input sample and then classify it into the class that has the highest frequency of labeled data in those Q clusters. Woolam et al. [10] used the label propagation technique instead of the nearest neighborhood algorithm.

Zhu et al. [11] proposed an active learning algorithm for selecting unlabeled samples which should be labeled in a chunk to build a new classifier for an ensemble. Their algorithm finds out one unlabeled sample with the largest ensemble variance from the entire set of unlabeled samples within a chunk. Such an unlabeled sample is manually labeled, and then the newly labeled sample is used together with existing labeled samples when the ensemble variance is calculated in the next iteration. This process is performed until there are no unlabeled samples in a chunk. The ensemble variance of an unlabeled sample is defined using errors for each classifier in an ensemble. Each classifier is evaluated on the set which consists of the current labeled samples within a chunk and the corresponding unlabeled sample with a predicted class.

The new ensemble method proposed in this paper is able to dynamically generate new classifiers for an ensemble on streaming unlabeled data without using "chunks". Our method decides if streaming samples should be selected for building new classifiers not according to a time interval, but according to a change in the distribution of streaming data. This leads to actively generate new classifiers for an ensemble. In particular, our approach is able to select the samples from streaming unlabeled data in an online process. We also introduce a new approach in classifying each input streaming sample.

This paper is organized as follows. The proposed ensemble method is described in Section 2. We report experimental results in Section 3 where our approach is compared with traditional methodologies and existing methods using real data sets including real streaming data sets, while conclusions works are presented in Section 4.

2   AEC: Active Ensemble Classifier

The chunk-based ensemble approach blindly builds new classifiers for an ensemble regardless of specific characteristics of data streams because it builds a new classifier periodically in a fixed interval of time. Obviously, this is an inefficient approach for real-world applications. Our ensemble approach decides dynamically when to build a new classifier and which samples should be used as a training data set for the new classifier according to changes in the distribution of streaming data.

2.1   Selecting Unlabeled Samples from Streaming Data for a New Classifier

Our ensemble approach builds the next new classifier of an ensemble using samples which do not belong to the current distribution of streaming data. If a sample belongs to the current distribution of streaming data, we may assume that the current ensemble correctly classifies the sample. Alternatively, if a sample belongs to a new data distribution different from the current one, the current ensemble might misclassify the sample. Our approach recognizes and selects samples which do not belong to the current distribution of streaming data, and the selected samples are used when forming a training data set for a new classifier.

We define the current distribution of streaming data using training data sets for the classifiers of the current ensemble. The current distribution (*CD*) of streaming data is represented as a set of pairs:

$$CD = \{(M_0, r_0), ..., (M_k, r_k)\} \tag{1}$$

where M_k and r_k are the mean vector (*center*) and the average distance of the training data set T_k used for the k-th classifier in the current ensemble, respectively. k is the number of classifiers in the current ensemble. Each pair represents a distribution of training data used to build a corresponding classifier in the ensemble. We have approximated a distribution of training data with an area in n-dimensionality space defined by a mean vector ($M = (m_1, m_2, ..., m_n)$), and an average distance (r) as radius. This area is referred as a *training-data area*.

$$m_i = \frac{1}{|T|} \sum_{j=1}^{|T|} x_{ji}, \quad X_j = (x_{j1}, x_{j2}, ..., x_{jn}) \in T \tag{2}$$

$$r = \frac{1}{|T|} \sum_{j=1}^{|T|} d(M, X_j) \tag{3}$$

where $dist(M, X_j)$ denotes distance function between a mean vector M and an input vector X_j. We employed the method used in [2] in order to calculate distance between data with categorical attributes.

We define samples which do not belong to the current distribution of streaming data as suspicious samples. If a sample X is not assigned to any of the training-data areas, it becomes a suspicious sample. $\varSigma$ denotes a set of suspicious samples.

$$\Sigma = \{X \mid d(M_i, X) > r_i \ \text{for all} \ i = 1, 2, ..., k\} \tag{4}$$

When forming a training data set for a new classifier, our ensemble method only chooses *neighboring suspicious* samples from Σ. Neighboring suspicious samples are able to generate a training-data area that is useful in accuracy of an ensemble. The neighboring suspicious samples are defined using a circle (or sphere) area such as a training-data area. Such an area will be referred to as the *neighbor area*. The center point of a neighbor area becomes a suspicious sample X_c ($\in \Sigma$) which does not belong to any of neighbor areas. This center is fixed until neighboring suspicious samples are transformed into training data. The size of a neighbor area is larger than one of the training-data area generated using the neighboring suspicious samples belonging to there. We have set the radius of a neighbor area as $2r_0$ (r_0 denotes the radius of the training-data area that is generated from an initial training data set). S denotes a set of neighboring suspicious samples.

$$S = \{X \mid d(X, X_c) < 2r_0, X \in \Sigma \ \text{and} \ X_c \in \Sigma\} \tag{5}$$

2.2 Classifying a New Streaming Sample

When an ensemble classifies a new sample, X, our ensemble approach selects a class with the largest weighted summation as shown in equation (6).

$$y = \arg\max{}_{y_k \in CLASS} \left(\sum_{j=1}^{n} w_j L_j (y_k) \right) \qquad (6)$$

$L_j(y_k)$ is the likelihood of the class y_k predicted by the j-th classifier of an ensemble. w_j is a weight value of the j-th classifier on the new samples X. n is the total number of classifiers in an ensemble, and $CLASS$ denotes the set of classes.

Whenever an ensemble classifies an input sample, the weight values of each classifier are calculated using the membership function of the fuzzy c-means algorithm as shown in equation (7). If the mean vector of a training-data area is the closest to a new sample, then the largest weight value is assigned to the corresponding classifier. The sum of classifier weights in an ensemble for each new sample is 1.

$$w_j = \left(\sum_{k=1}^{n} \left(\left(\frac{d(X,M_j)}{r_j} \right) \middle/ \left(\frac{d(X,M_k)}{r_k} \right) \right)^2 \right)^{-1} \qquad (7)$$

w_j is a weight value of the j-th classifier of an ensemble on a new sample X. M_j and r_j denote the mean vector and the radius of the training-data area representing the j-th classifier of an ensemble. n is the total number of classifiers of the ensemble. $d(X,M_j)$ denotes the distance between a new sample vector X and the mean vector M_j of the training-data area of the j-th classifier. We use normalized distance dividing a distance $d(X,M_j)$ by the corresponding average distance r_j because the sizes of training-data areas are different from each other.

3   Experiments

We chose seven large real data sets from the UCI data repository and three streaming data sets from Wikipedia under the keyword "concept drift", as shown in Table 1. These sets have various numbers of classes and various types of attributes. For example, the "Adult" data set consists of six numerical attributes and nine categorical attributes including a class attribute. Some classes are represented with very small number of samples; the "Nursery" data set has only two samples belonging to the 2nd class ("Recommend"). There is no sample with the 2nd class in the initial training data set as shown in Table 1.

The original "Ozone" and "Electricity market" data sets have samples with missing values for some numerical attributes. We removed those samples from the original data sets. The "Mushroom" and "Adult" data sets also have missing values for some categorical attributes. We replaced these missing values in each sample with the new, categorical "NULL" value.

We divided each data set into an initial training data set and a streaming data set as shown in Table 1. The initial training data set was used for building an initial classifier of an ensemble, the streaming data set was used as a test data set for evaluating ensemble approaches. The first 1,000 samples (15.5% of the "Landsat Satellite, 12.3% of the "Mushroom", 7.7% of the "Nursery", 5.25% of the "MAGIC Gamma Telescope", 2.04% of the "Adult", 0.17% of the "Covertype") was used as the initial training data. The remaining samples were used as streaming data. Since the

"Ozone" set has only 1,848 samples, the first 200 samples were used as the initial training data. For the three streaming data sets ("Electricity Market", "PAKDD2009", and "KDDCup1999"), we used the first 1,600 samples as the initial training data to compare with the approach proposed by Woolam et al.[10] where the chunk size was defined as 1,600 consecutive sequence samples in their experiments with the "KDDCup1999" data set. Table 1 also shows the class distributions for the initial training data and the rest of streaming data. In the initial training data sets for the "Nursery", and "Covertype" data sets, samples of all classes are not represented. We selected these data sets to verify the performances of the proposed methodology in a case when all classes are not defined in the model initially.

Table 1. Class rates for the initial training data and the rest of streaming data

Data	#Attributes	Initial training data		Streaming data	
		#data	Rate of Each Class (%)	#data	Rate of Each Class (%)
Ozone[1]	73	200	C1(95.0),C2(5.0)	1,648	C1(97.1),C2(2.9)
Landsat Satellite[1]	37	1,000	C1(0.0),C2(21.7), C3(50.8),C4(7.4), C5(6.0),C6(14.1)	5,435	C1(28.2),C2(8.9), C3(15.6),C4(10.2), C5(11.9),C6(25.2)
Mushroom[1]	23	1,000	C1(89.8),C2(10.2)	7,124	C1(46.5),C2(53.5)
Nursery[1]	10	1,000	C1(32.8),C2(0.0), C3(2.3),C4(34.9), C5(30.0)	11,960	C1(33.4),C2(0.015), C3(2.6),C4(32.7), C5(31.3)
MAGIC[*1]	11	1,000	C1(65.1), C2(34.9)	18,020	C1(64.8),C2(35.2)
Adult[1]	15	1,000	C1(23.2), C2(76.8)	47,842	C1(23.9),C2(76.1)
Covertype[1]	55	1,000	C1(22.6), C2(58.5), C3(0.0),C4(0.0),C5(18.9), C6(0.0),C7(0.0)	580,012	C1(36.5),C2(48.7), C3(6.2),C4(0.5),C5(1.6), C6(3.0),C7(3.5)
EM[*2]	7	1,600	C1(39.8),C2(60.2)	25,949	C1(41.7),C2(58.3)
PAKDD[*2]	24	1,600	C1(79.1), C2(20.9)	48,400	C1(80.3),C2(19.7)
KDDCup[*2]	42	1,600	C1(99.9), C2(0.1)	492,421	C1(19.4),C2(80.6)

*MAGIC: "MAGIC Gamma Telescope", EM: "Electricity Market", PAKDD: PAKDD2009, KDDCup:KDDCup1999 (1: UCI repository, 2: Wikipedia under the keyword "Concept drift")

We used the following three performance measures for evaluation of our system and comparison with other traditional ensemble methodologies for streaming data:

The total number of new classifiers (TC) is a count of generated new classifiers for an ensemble over streaming data (after manually labeling samples in an offline process). *The labeled sample rate (LR)* is the proportion of the labeled samples used for building new classifiers for an ensemble in a data stream.

The weighted sum of F-measures for all classes (WSF) is an appropriate measure for an ensemble accuracy applied for streaming data in a multi-classification problem with skewed class distribution. This measure is defined as follows:

$$WSF = \sum_{c_i \in CLASS} w_i F(c_i) \tag{8}$$

where *CLASS* denotes the set of classes, and $F(c_i)$ is the harmonic mean of the precision and the recall of c_i class. The weight w_i of a class c_i is determined using equation (9), where N is the number of streaming data, and n_i is the number of samples belonging to c_i class in the streaming data.

$$w_i = \frac{1}{|CLASS|-1} \times (1 - \frac{n_i}{N})$$ (9)

A weight value according to the proportion of a class was divided by |*CLASS*|-1 so that the sum of weights of the classes becomes normalized to 1.

3.1 Comparison with the Chunk-Based Ensemble Approach

We implemented simple voting ensemble (SVE) and weighted ensemble (WE) methods to show that our method efficiently maintains performance of an ensemble. Both the SVE and WE methods periodically build new classifiers for an ensemble using samples that are randomly selected from samples within each chunk. We defined the chunk size as 1,600 samples and built a new classifier using 10% labeled samples (160 samples), as in work by Woolam et al.[10]. The SVE method combined results of classifiers in an ensemble by the majority voting method. The WE method used the weighted majority voting method as the combining method for classification. In the WE method, each classifier weight was determined using the most recent chunk according to the method presented by Wang et al. [12]. If a classifier in an ensemble provides the highest accuracy on the samples of the most recent chunk, the largest value will be attached to the classifier weight. We defined the minimum number of the neighboring suspicious samples within a neighbor area in our method as 70 for the "Ozone" data set and 300 for the others.

Table 2. Comparison with TC, LR and WSF of SVE and WE

Data set	AEC			SVE			WE		
	TC	LR (%)	WSF	TC	LR (%)	WSF	TC	LR (%)	WSF
Ozone	2	8	0.145	1	10	0.136 ±0.0001	1	10	0.136 ±0.0005
Landsat Satellite	2	11	0.501	3	9	0.532±0.03	3	09	0.525±0.03
Mushroom	8	33	0.887	4	9	0.561±0.11	4	09	0.742±0.06
Nursery	4	10	0.485	7	9	0.442±0.01	7	09	0.460±0.02
MAGIC	6	9	0.774	11	10	0.740±0.01	11	10	0.692±0.02
Adult	7	4	0.510	29	10	0.549±0.06	29	10	0.529±0.03
Covertype	31	1	0.328	362	10	0.215±0.02	362	10	0.117±0.01
EM	6	12	0.643	17	10	0.636±0.01	17	10	0.601±0.01
PAKDD	9	8	0.290	31	10	0.187±0.01	31	10	0.351±0.01
KDDCup	25	2	0.978	308	10	0.669±0.001	308	10	0.701±0.01
AVERAGE	10.0	9.8	0.554	77.3	9.7	0.466	77.3	9.7	0.485

The SVE and WE methods produced the same total number of new classifiers (TC) and the same labeled sample rate (LR) as shown in Table 2 because they use the chunk-based approach. The labeled sample rate is the proportion of samples that are used for building new classifiers. If the number of samples within the last chunk does not reach a chunk size, the correct labels of the samples are not used to build a new classifier. That is the reason that the values of their LR in Table 2 do not become 10%. We performed ten experiments for both SVE and WE with different random seeds for selecting 160 samples to be labeled in each chunk. Our ensemble, AEC (Active Ensemble Classifier), generated 87.1% fewer new classifiers than the chunk-based ensemble approach using partially labeled samples, and used an average of 10% labeled samples for the ten data sets. AEC produced an average of 0.554. This average is 18.8% higher than the average WSF of SVE. This difference in WSF is statistically significant (Wilcoxon's test, significant level=0.05, T =8 is equal to the critical value=8).

3.2 Comparison with Existing Ensemble Methods Using Semi-supervised Learning

The results of our ensemble method compared with results of Masud's method [9], and Woolam's method [10]. Those existing methods construct an ensemble using few labeled samples within each chunk, which consists of consecutive sequence samples. Masud's and Woolam's methods used their semi-supervised learning algorithm to build a new classifier from all samples within a chunk.

Table 3. Comparison with Masud's and Woolam's methods using semi-supervised learning

Ensemble method	Accuracy (%)	TC	LR(%)
AEC	99.04	25	2
Woolam's method with Bias [10]	97.69	308	10
Woolam's method with Random [10]	98.06	308	10
Masud's method [9]	92.57	308	10

Table 3 shows the results of AEC, Masud's and Woolam's methods for the "KDDCup1999" data set. The accuracy values of Masud's and Woolam's methods in Table 3 were derived from Woolam et al [10]. In their experiment, they predefined the chunk size as 1,600 samples. They randomly selected 160 samples (10%) from a chunk, and then used the samples' correct classes for their semi-supervised learning. Accordingly, we can calculate TC as 308 (494,021 / 1,600 = 308,763), and the number of labeled samples as 49,280 (160 * 308 = 49,280). AEC generated 91.8% fewer new classifiers, and used about 80% fewer labeled samples than the two existing methods. Our ensemble method even produces an average of 1.17% higher classification accuracy than two types of Woolam's method, and 6.98% higher than Masud's method.

3.3   Comparison among AEC Using Different Classification Algorithms

We used decision trees as classifiers of an ensemble for the experiments so far. The decision tree was generated with J48 decision tree (C4.5 algorithm) from Weka(www.cs.waikato.ac.nz/ml/weka/). However, our ensemble approach does not depend on a specific classification algorithm for building a classifier of an ensemble. To prove that, we carried out experiments on ten real data sets with three other classification algorithms: SVM(Support Vector Machine), MLP(Multilayer Perceptron), and NB(Naïve Bayesian). The three kinds of classifier were also built with each algorithm provided by Weka.

Table 4 shows WSF values produced when each algorithm was used. To use the Friedman Test as in Demšar [12] the algorithms achieved their ranks according to WSF values for each data set separately. Numbers in parentheses in Table 4 denote ranks of algorithms. In case of ties (like Lansat Satellite, and KDDCup1999), average ranks are assigned. In this test, the null-hypothesis is that all the classifiers perform the same and the observed differences are merely random. With four algorithms and ten data sets, FF=1.93 is distributed according to the F distribution with 4-1=3 and (4-1)×(10-1)=27 degree of freedom. The critical value of $F(3,27)$ for $\alpha=0.05$ is 2.96, so we accept the null-hypothesis.

Table 4. Comparison of WSF for AECs with each of four classification algorithms: Decision Tree(DT), SVM, MLP, and NB; Numbers in the parentheses denote ranks of algorithms

Data Set	AEC with DT	AEC with SVM	AEC with MLP	AEC with NB
Ozone	0.145 (3)	0.028 (4)	0.159 (2)	0.164 (1)
Landsat Satellite	0.501 (4)	0.505 (3)	0.544 (1.5)	0.544 (1.5)
Mushroom	0.887 (4)	0.933 (2)	0.931 (3)	0.938 (1)
Nursery	0.485 (2)	0.563 (1)	0.481 (3)	0.465 (4)
MAGIC Gamma Telescope	0.774 (2)	0.732 (3)	0.800 (1)	0.661 (4)
Adult	0.510 (4)	0.654 (1)	0.647 (3)	0.648 (2)
Covertype	0.328 (4)	0.351 (3)	0.362 (2)	0.374 (1)
Electricity Market	0.643 (3)	0.651 (2)	0.641 (4)	0.658 (1)
PAKDD2009	0.290 (3)	0.177 (4)	0.332 (2)	0.439 (1)
KDDCup1999	0.978 (2.5)	0.954 (4)	0.984 (1)	0.978 (2.5)
Average rank	3.15	2.7	2.25	1.9

3.4   Comparison with an Online Learning Approach

We compare WSF values of AEC and an online learning approach which is another methodology for streaming data classification. The online learning approach refines a single classifier using correct classes for all streaming data. In other words, after predicting the true class of an input streaming sample, a single classifier improves by using its correct class. In this experiment, we used the Naïve Bayesian algorithm provided by Weka as a classification algorithm because it is designed to facilitate online learning as well as batch learning.

Table 5. Comparison of WSF for AEC with Naïve Bayesian and online Naïve Bayesian classifier

Data set	AEC with NB	online Naïve Bayesian classifier
Ozone	0.164	0.510
Landsat Satellite	0.544	0.753
Mushroom	0.938	0.981
Nursery	0.465	0485
MAGIC Gamma Telescope	0.661	0.609
Adult	0.648	0.671
Covertype	0.374	0.307
Electricity Market	0.658	0.595
PAKDD2009	0.439	0.409
KDDCup1999	0.978	0.966
AVERAGE	0.586	0.592

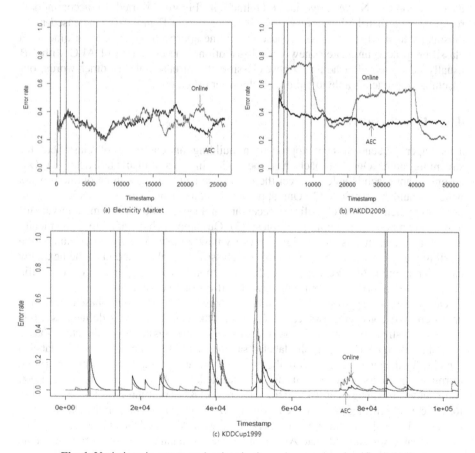

Fig. 1. Variations in errors, and points in time where a new classifier is built

Table 5 shows the WSF values of AEC with NB (Naïve Bayesian) and the online Naïve Bayesian method. AEC with NB produced comparable WSF with the online

Naïve Bayesian method in all the experiments (Wilcoxon's test, significant level=0.05, T=24 > critical value = 8).

We compared our approach and the online learning approach in a time domain with the three real streaming data sets: "Electricity Market", "PAKDD2009", and "KDDCup1999." Figure 1 shows variations in prequencial errors of AEC with NB and the online Naïve Bayesian method. As in work by Gama et al.[14], the prequencial error is calculated by a forgetting mechanism using fading factors. The vertical lines in each chart of Figure 1 shows points in time where a new classifier for an ensemble is built. Figure 1-(c) shows variations of errors occurred between the 1st and the 100,000th network traffic streaming samples where the "ATTACK" class is increasing in frequency.

The error rate of a classifier in online learning approaches increases until the classifier adapts to a new or recurrent data distribution. If an online classifier adapts to a coming data distribution, its error rate will decrease. In the PAKDD2009 streaming data, the online Naïve Bayesian ("Online" in Figure 1) tried to accommodate recurrent data distributions every time. Its error rate in Figure 1-(c) increased more drastically than AEC with NB because the online approach requests more time for a classifier to accommodate a new data distribution. The error rate of AEC with NB usually decreased after adding a new classifier to an ensemble. In other words, our ensemble approach is able to build new classifier efficiently.

4   Conclusions

This paper presents a new approach in building an ensemble of classifiers for streaming data. Our approach has the following main characteristics: (1) Our ensemble approach is able to select the most promising samples in an online process which should be labeled; (2) Our approach is able to build new classifiers for an ensemble when the new classifier is necessary, not systematically in time intervals for a fixed number of streaming samples; (3) Our approach is able to dynamically accommodate each classifier weight for every new sample to be classified, unlike the existing methods where an ensemble keeps classifier weights fixed until the next new classifier is built. (4) We confirmed that our approach is independent of a specific classification algorithm for building new classifiers of an ensemble.

Our ensemble approach, AEC, was compared with the chunk-based ensemble approach. We considered two cases. (1) When the chunk size was defined as 1,600 consecutive streaming data, AEC generated 12.9% new classifiers for the chunk-based ensemble approach using partially labeled samples, and used an average of 10% labeled samples for the ten data sets. (2) In the existing ensemble methods with semi-supervised learning, classifiers were built using 10% labeled samples in every chunk of 1,600 consecutive streaming data. AEC built 8.1% new classifiers using 20% labeled samples for the existing method. In all the experiments, AEC produced comparable classification accuracy. Through comparison of the performance for AEC, and the online approach in a time domain, we showed that AEC can efficiently maintain the performance of an ensemble over streaming data.

We are planning to apply our ensemble approach to a system which detects click fraud from click streaming data in online advertising networks. We expect that dynamics of new fraud types may be detected online with our approach.

Acknowledgements. This work was supported by the National Research Foundation of Korea Grant funded by the Korean Government (Ministry of Education, Science and Technology). [NRF-2010-357-D00209]

References

1. Minku, L.L., Yao, X.: DDD: A New Ensemble Approach for Dealing with Concept Drift. IEEE Transactions on Knowledge and Data Engineering (99) (2011), doi:10.1109/TKDE.2011.58
2. Ryu, J.W., Kantardzic, M., Walgampaya, C.: Ensemble Classifier Based on Misclassified Streaming Data. In: Proc. of the 10th IASTED Int. Conf. on Artificial Intelligence and Applications, Austria, pp. 347–354 (2010)
3. Gao, J., Fan, W., Han, J.: On Appropriate Assumptions to Mine Data Streams: Analysis and Practice. In: Proc. of the 7th IEEE ICDM, USA, pp. 143–152 (2007)
4. Wang, H., Fan, W., Yu, P.S., Han, J.: Mining Concept-Drifting Data Streams using Ensemble Classifiers. In: Proc. of the 9th ACM SIGKDD KDD, USA, pp. 226–235 (2003)
5. Chu, F., Zaniolo, C.: Fast and Light Boosting for Adaptive Mining of Data Streams. In: Dai, H., Srikant, R., Zhang, C. (eds.) PAKDD 2004. LNCS (LNAI), vol. 3056, pp. 282–292. Springer, Heidelberg (2004)
6. Zhang, P., Zhu, X., Shi, Y.: Categorizing and Mining Concept Drifting Data Streams. In: Proc. of the 14th ACM SIGKDD, USA, pp. 812–820 (2008)
7. Zhang, P., Zhu, X., Shi, Y., Wu, X.: An Aggregate Ensemble for Mining Concept Drifting Data Streams with Noise. In: Theeramunkong, T., Kijsirikul, B., Cercone, N., Ho, T.-B. (eds.) PAKDD 2009. LNCS (LNAI), vol. 5476, pp. 1021–1029. Springer, Heidelberg (2009)
8. Wei, Q., Yang, Z., Junping, Z., Youg, W.: Mining Multi-Label Concept-Drifting Data Streams Using Ensemble Classifiers. In: Proc. of the 6th FSKD, China, pp. 275–279 (2009)
9. Masud, M.M., Gao, J., Khan, L., Han, J., Thuraisingham, B.: A Practical Approach to Classify Evolving Data Streams: Training with Limited Amount of Labeled Data. In: ICDM, Pisa, Italy, pp. 929–934 (2008)
10. Woolam, C., Masud, M.M., Khan, L.: Lacking Labels in the Stream: Classifying Evolving Stream Data with Few Labels. In: Rauch, J., Raś, Z.W., Berka, P., Elomaa, T. (eds.) ISMIS 2009. LNCS, vol. 5722, pp. 552–562. Springer, Heidelberg (2009)
11. Zhu, X., Zhang, P., Lin, X., Shi, Y.: Active Learning from Data Streams. In: Proceeding of the 7th IEEE International Conference on Data Mining, USA, pp. 757–762 (2007)
12. Wang, H., Fan, W., Yu, P.S., Han, J.: Mining Concept-Drifting Data Streams using Ensemble Classifiers. In: Proc. of the 9th ACM SIGKDD, USA, pp. 226–235 (2003)
13. Demšar, J.: Statistical Comparisons of Classifiers over Multiple Data Sets. Journal of Machine Learning Research 7, 1–30 (2006)
14. Gama, J., Sebastião, R., Rodrigues, P.P.: Issues in Evaluation of Stream Learning Algorithms. In: Proceeding of the 15th ACM SIGKDD International Conference on Knowledge Discovery in Data Mining, France, pp. 329–338 (2009)

I/O Efficient Algorithms for Block Hessenberg Reduction Using Panel Approach

Sraban Kumar Mohanty[1] and Gopalan Sajith[2]

[1] Computer Science & Engineering Discipline,
PDPM Indian Institute of Information Technology,
Design and Manufacturing Jabalpur, Jabalpur-482005, MP, India
sraban@iiitdmj.ac.in
[2] Computer Science & Engineering Department, Indian Institute of Technology
Guwahati, Guwahati-781039, Assam, India
sajith@iitg.ac.in

Abstract. Reduction to Hessenberg form is a major performance bottleneck in the computation of the eigenvalues of a nonsymmetric matrix; which takes $O(N^3)$ flops. All the known blocked and unblocked direct Hessenberg reduction algorithms have an I/O complexity of $O(N^3/B)$. To improve the performance by incorporating matrix-matrix operations in the computation, usually the Hessenberg reduction is computed in two steps: the first reducing the matrix to a banded Hessenberg form, and the second further reducing it to Hessenberg form. We propose and analyse the first step of the reduction, i.e., reduction of a nonsymmetric matrix to banded Hessenberg form of bandwidth t for varying values of N and M (the size of the internal memory), on external memory model introduced by Aggarwal and Vitter for the I/O complexity and show that the reduction can be performed in $O(N^3/\min\{t, \sqrt{M}\}B)$ I/Os.

Keywords: Large Matrix Computation, External Memory Algorithms, Out-of-Core Algorithms, Matrix Computations, Hessenberg Reduction, I/O Efficient Eigenvalue Problem.

1 Introduction

In traditional algorithm design, it is assumed that the main memory is infinite in size and allows random uniform access to all its locations. This enables the designer to assume that all the data fits in the main memory. (Thus, traditional algorithms are often called "in-core".) Under these assumptions, the performance of an algorithm is decided by the number of instructions executed.

These assumptions may not be valid while dealing with massive data sets, because in reality, the main memory is limited, and so the bulk of the data may have to be stored in slow secondary memory. The number of inputs/outputs (I/Os) executed would be a more appropriate performance metric in this case.

In the last few decades, computers have become a lot faster, and the amount of main memory they have has grown. But the issue of the main memory being limited has only become more relevant, because applications have grown even

S. Srinivasa and V. Bhatnagar (Eds.): BDA 2012, LNCS 7678, pp. 134–147, 2012.

faster in size. Also, small computing devices (e.g., sensors, smart phones) with limited memories have found several uses.

Designing of I/O efficient algorithms (also called out-of-core (OOC) algorithms) has, therefore, been an actively researched area in the last few years. For many a problem, it has been shown that the traditionally efficient algorithm is not very I/O efficient, and that novel and very different design techniques can be used to produce a much more I/O efficient algorithm. The External Memory model of Aggarwal and Vitter [1, 2] has been used to design many of these algorithms. This model has a single processor and a two level memory. It is assumed that the bulk of the data is kept in the secondary memory (disk) which is a permanent storage. The secondary memory is divided into blocks. An I/O is defined as the transfer of a block of data between the secondary memory and a volatile main memory, which is limited in size. The processor's clock period and main memory access time are negligible when compared to secondary memory access time. The measure of performance of an algorithm is the number of I/Os it performs. The model defines the following parameters: the size of the main memory (M), and the size of a disk block $(B < M)$. (The two levels might as well be the cache and the main memory.) [3, 4]

Early work on External Memory algorithms were largely concentrated on fundamental problems like sorting, permutating, graph problems, computational geometry and string matching problems [1, 5–12]. External Memory algorithms for fundamental matrix operations like matrix multiplication and matrix transposition were proposed in [1, 12]. Not many linear algebra algorithms have been designed or analysed explicitly on the External Memory model [3].

Software libraries like BLAS, LAPACK, ScaLAPACK, PLAPACK, POOCLA-PACK, SOLAR [13–19] that implement various linear algebra operations have been in development over many years. See [3, 20, 21] for a survey on out-of-core algorithms in linear algebra.

Computing of eigenvalues of matrices have wide-ranging applications in engineering and computational sciences such as control theory, vibration analysis, electric circuits, signal processing, pattern recognition, numerical weather prediction and information technology, to name a few [22–25].

The eigenvalues are typically computed in a two stage process. In the first stage the input matrix is reduced, through a sequence of orthogonal similarity transformations (OSTs) to a condensed form and in the second stage an iterative method called the QR algorithm [4] is applied to the condensed form. (An OST on a square matrix $A \in \mathbb{R}^{N \times N}$ transforms A into $Q^T A Q$ using an orthogonal matrix Q. OSTs preserve eigenvalues.) All known algorithms for the first stage need $O(N^3/B)$ I/Os and $O(N^3)$ flops [3]. Therefore, the first stage is usually the performance bottleneck.

The traditional Hessenberg reduction algorithms based on Householder transformations (Householders) or rotators are rich in vector-vector (V-V) and vector-matrix (V-M) operations, and not in matrix-matrix (M-M) multiplications [26, 27]. Most of the computations involving V-V or V-M operations, perform

$O(N^2/B)$ I/Os while executing $O(N^2)$ flops thus giving the so called *surface-to-surface* effect to the ratio of flops to I/Os [26, 27]. It has been observed that casting computations in terms of M-M operations improves performance by reducing I/Os [3, 26, 27]. The reason is that M-M operations perform only $O(N^2/B)$ I/Os while executing $O(N^3)$ flops giving the so called *surface-to-volume* effect to the ratio of flops to I/Os [28].

Algorithms can be recast to be rich in M-M operations using either the *slab* or the *tile* approach. In the former, the matrix is partitioned into vertical slabs, which are processed one after the other; between two slabs the whole of the matrix may have to be updated using an aggregate of transformations generated from the processing of the slab; this updation could be done using M-M operations. Slab based algorithms for Cholesky, LU and QR decompositions, and for QR and QZ algorithms, that use aggregations of Gauss elimination or Householder transformations have been reported [4, 29–31]. In the tile approach, the matrix is typically partitioned into square tiles [13, 14, 32, 33]. It has been observed that the tile approach results in more scalable out-of-core algorithms for Cholesky [13, 14, 32] and QR decompositions [33], and in more granular and asynchronous parallel algorithms on multicore architectures for LU, Cholesky, and QR decompositions [34].

We focus on Householder transformation based reductions, because these exploit locality of reference better than rotation based algorithms [35], and therefore are more amenable to OOC computations. (If $u \in \mathbb{R}^N$, then there exists a symmetric orthogonal matrix $Q_u = (I - \beta u u^T) \in \mathbb{R}^{N \times N}$, $\beta = 2/\|u\|_2$, called the Householder of u, such that $Q_u u = -u$ and, for any vector v orthogonal to u, $Q_u v = v$.) Householder based sequential and parallel algorithms for Hessenberg reduction using the slab approach have been proposed [36–38]. But even in these, V-M operations dominate the performance: after the reduction of each column of a slab, the rest of the matrix needs to be read in to update the next column before it will be ready to be reduced. This is because of the two sidedness of the transformations involved. This also makes it difficult to design tile based versions of these algorithms. To the best of our knowledge, no tile based algorithm has been proposed for the above direct reductions.

Due to the above reasons, it has been proposed that the reduction in the first stage be split into two steps [3, 35, 39–42], the first reducing the matrix to a block condensed form (banded Hessenberg) [43–45], and the second further reducing it to Hessenberg form [35, 40, 41, 46]. Almost all operations of the first step are performed using M-M operations. Though the modified first stage takes more flops, its richness in M-M operations makes it more efficient on machines with multiple levels of memory. Usually these reductions are implemented using relatively expensive two sided orthogonal transformations rather than the inexpensive single sided Gauss elimination, because orthogonal transformations guarantee stability [27, 47].

In this paper, we study the first step of the first stage reduction and show that it takes $O(N^3/tB)$ I/Os if $t \leq \sqrt{M}$ and $O(N^3/\sqrt{M}B)$ I/Os if $t > \sqrt{M}$ [3]. We now present an overview of the results in this paper.

1.1 Reduction of a Nonsymmetric Matrix to Banded Hessenberg Form

Suppose the input is a nonsymmetric matrix $A \in \mathbb{R}^{N \times N}$, and it is to be reduced to banded upper Hessenberg form H_t of bandwidth t. This could be done using an OST: construct an orthogonal matrix $Q \in \mathbb{R}^{N \times N}$ such that $H_t = Q^T A Q$ [26, 42, 43, 48]. A slab based sequential algorithm [26], and a parallel algorithm for message passing multicomputers [42] are known. Tile based algorithms for multicore architectures using Householders and Givens rotations are presented in [43, 48].

We study the slab based algorithm of [26], and analyse it for its I/O complexity for varying values of M and N, for a given value of t. This algorithm assumes that the slab size $k = t$. We generalise this into an algorithm with k not necessarily the same as t. We find that the algorithm performs the best when $k = \min\{t, \sqrt{M}\}$.

1.2 Organisation of This Paper

In Section 2, we propose and analyse the panel based reduction algorithm to reduce a nonsymmetric matrix to banded Hessenberg form.

2 Reduction of a Nonsymmetric Matrix to Banded Hessenberg Form Using the Slab Approach

Let $A \in \mathbb{R}^{N \times N}$ be the nonsymmetric matrix that is to be reduced to banded upper Hessenberg form H_t of bandwidth t.

2.1 Reduction with a Slab Width of t

The algorithm of [26], when $N - t > t$, partitions the matrix into slabs of width t, and then proceeds in $N/t - 1$ iterations, each of which reduces a slab using block QR factorisations and then updates the rest of the matrix from both sides using aggregated Householders [49]. The algorithm is illustrated below:

Consider the first slab of A:

$$A = \begin{array}{c} \\ \\ \end{array} \begin{bmatrix} \overset{t}{A_{11}} & \overset{N-t}{A_{12}} \\ \hline A_{21} & A_{22} \end{bmatrix} \begin{array}{c} t \\ \\ N-t \end{array}$$

Perform a QR factorisation on A_{21} such that $A_{21} = Q_{21} R_{21}$. Then an OST with $Q = \begin{pmatrix} I_t & 0 \\ 0 & Q_{21} \end{pmatrix}$ gives:

$$Q^T A Q = \begin{pmatrix} A_{11} & A_{12} Q_{21} \\ \hline R_{21} & Q_{21}^T A_{22} Q_{21} \end{pmatrix}$$

If Q is in WY-representation [49], this involves computing a product of the form $(I + WY^T)^T A(I + WY^T)$. Now the first slab is in banded Hessenberg form. Repeat this $N/t - 1$ times, and the matrix reduces to banded upper Hessenberg form H_t:

$$H_t = \begin{pmatrix} H_{11} & H_{12} & \cdots & & \cdots & H_{1\frac{N}{t}} \\ H_{21} & H_{22} & \cdots & & \cdots & H_{2\frac{N}{t}} \\ 0 & H_{32} & \ddots & & \cdots & H_{3\frac{N}{t}} \\ \vdots & \vdots & \ddots & \ddots & & \vdots \\ 0 & 0 & \cdots & H_{\frac{N}{t}(\frac{N}{t}-1)} & & H_{\frac{N}{t}\frac{N}{t}} \end{pmatrix}$$

where each H_{ij} is a $t \times t$ tile; H_{ij} is a full tile when $i \leq j$, upper triangular when $i = j + 1$, and all zero otherwise. Thus, H_t is banded upper Hessenberg with bandwidth t.

The I/O complexity of this algorithm can obtained from that of Case-1 in Algorithm 2.1, by substituting $k = t$.

If $N - t \leq t$ then A is partitioned as follows:

$$A = \begin{array}{cc} & \begin{array}{cc} N-t & t \end{array} \\ \left[\begin{array}{c|c} A_{11} & A_{12} \\ \hline A_{21} & A_{22} \end{array}\right] & \begin{array}{c} t \\ N-t \end{array} \end{array}$$

Perform a QR factorisation on A_{21} such that $A_{21} = Q_{21} R_{21}$. Then an OST with $Q = \begin{pmatrix} I_t & 0 \\ 0 & Q_{21} \end{pmatrix}$ gives $Q^T A Q$ which is in the desired banded Hessenberg form. The I/O complexity of this is as given in Table 1.

Table 1. The case of $(N - t) < t$. The number of I/Os for QR decomposition, matrix multiplication, and the total are given in columns titled QRD, MM and Total respectively.

Conditions on M & $N - t$	QRD	MM	Total
$N - t > M$	$\frac{(N-t)^3}{\sqrt{M}B}$	$\frac{N(N-t)^2}{\sqrt{M}B}$	$\frac{N(N-t)^2}{\sqrt{M}B}$
$N - t \leq \sqrt{M}$	$\frac{(N-t)^2}{B}$	$\frac{N(N-t)}{B}$	$\frac{N(N-t)}{B}$
$M^{2/3} < N - t \leq M$	$\frac{(N-t)^3}{\sqrt{N-t}B}$	$\frac{N(N-t)^2}{\sqrt{M}B}$	$\frac{(N-t)^3}{\sqrt{N-t}B} + \frac{N(N-t)^2}{\sqrt{M}B}$
$\sqrt{M} < N - t \leq M^{2/3}$	$\frac{(N-t)^4}{MB}$	$\frac{N(N-t)^2}{\sqrt{M}B}$	$\frac{(N-t)^4}{MB} + \frac{N(N-t)^2}{\sqrt{M}B}$

Algorithm 2.1. *Banded Hessenberg reduction using the slab approach*

Input: *An $N \times N$ matrix A, bandwidth t and algorithmic block size k.*
Output: $H = Q^T A Q$, *H is $N \times N$ banded upper Hessenberg with bandwidth t, a number of Householder aggregates with Q as their product.*

Case-1: $k \leq t$. *For ease of exposition, assume that k divides t*
  for $i = 1$ **to** $(N - t)/t$ **do**
    Let $g = (i - 1)t$
    for $j = 1$ **to** t/k **do**
      Let $h = (j - 1)k$
      QR-Decompose($A[g + h + t + 1 : N, \; g + h + 1 : g + h + k]$)
      Let $Q_{ij} = (I + Y_{ij} T_{ij} Y_{ij}^T)$ be the resulting compact WY representation of Q
      Update the rest of the matrix:
      Left-multiply $A[g + h + t + 1 : N, \; g + h + k + 1 : N]$ with Q_{ij}
      Right-multiply $A[1 : N; \; g + h + t + 1 : N]$ with Q_{ij}
    endfor
  endfor
Case-2: $k > t$. *For ease of exposition, assume that t divides k;*
  for $i = 1$ **to** $(N - t)/k$ **do**
    Let $g = (i - 1)k$
    Let y denote the range $g + t + 1 : N$
    Let z denote the range $g + k + 1 : N$
    $\hat{A}[y, \; y] = A[y, \; y]$
    for $j = 1$ **to** k/t **do**
      Let $h = (j - 1)t$
      QR-Decompose($A[g + h + t + 1 : N, \; g + h + 1 : g + h + t]$)
      Let $Q_{ij} = (I + Y_{ij} T_{ij} Y_{ij}^T)$ be the resulting compact WY representation of Q
      Let $Y_i = Y_{ij}$ if $j = 1$ and $Y_i = (Y_i \; Y_{ij})$ otherwise
      Let $T_i = T_{ij}$ if $j = 1$ and $T_i = \begin{pmatrix} T_i & T_i Y_i^T Y_{ij} T_{ij} \\ 0 & T_{ij} \end{pmatrix}$ otherwise
      Let x denote the range $g + h + t + 1 : g + h + 2t$
      If $(j \neq k/t)$ **do**
        Update the next t columns of the panel i:
        Compute $B_{ij} = \hat{A}[y, \; x] + \hat{A} Y_i T_i \left(Y_i^T [1 : h + t, \; x] \right)$
        Compute $A[y, \; x] = (I + Y_i T_i Y_i^T) B_{ij}$
      end if
    end for
    Update the rest of the matrix:
    Right-multiply $A[1 : g + t, \; y]$ with $(I + Y_i T_i Y_i^T)$
    Compute $A[y : z] = \hat{A}[y : z] + \hat{A} Y_i T_i \left(Y_i^T [1 : k, z] \right)$
    Left-multiply $A[y : z]$ with $(I + Y_i T_i Y_i^T)$
  endfor

2.2   Reduction with a Slab Width Not Necessarily Equal to t

In the above algorithm, if the slabs to be QR decomposed are very small compared to the main memory $((N - it) \times t \ll M)$, then the main memory would seem under utilised; a larger slab width might be appropriate. Also, if the slab is too big to fit in the main memory $((N - it) \times t \gg M$ then the QR factorisations and subsequent updates are to performed out-of-core; a smaller slab width might do better. We present an out-of-core algorithm (see Algorithm 2.1) that uses a slab width of k not necessarily the same as t.

Algorithms of a similar vein have been proposed for the reduction of a full symmetric matrix to symmetric banded form [44, 50, 51], and it has been observed [51] that choosing a value of at most t for k will do better than otherwise. To the best of our knowledge, no such out-of-core algorithm has been proposed for banded Hessenberg reduction. Case-2 of Algorithm 2.1 has been inspired by the parallel algorithm of [50], while Case-1 is a generalisation of the algorithm described in Subsection 2.1 [26].

Algorithm 2.1 divides A into vertical slabs of k columns each. Given the values of M, N and t, the I/O complexity of the algorithm depends upon the choice of k. We analyse the two cases of $t < k$ and $t \geq k$ separately, with an intent of finding the choice of k that would minimise the number of I/Os.

Case 1: $k \leq t$ For $1 \leq i \leq (N - t)/t$ and $1 \leq j \leq t/k$, in the (i,j)-th iteration there are a QR decomposition of an $(N - g - h - t) \times k$ matrix, a matrix multiplication chain of dimensions $(N - g - h - t, k, k, N - g - h - t, N - g - h - k)$, and a matrix multiplication chain of dimensions $(N, N - g - h - t, k, k, N - g - h - t)$, where $g = (i - 1)t$ and $h = (j - 1)k$. Let α, β and γ denote the I/O complexities of these operations, in that order. Clearly, $\gamma \geq \beta$, in all cases. As i varies from 1 to $(N - t)/t$ and j varies from 1 to t/k, $(N - g - h - t)$ takes on values lk for $(N - t)/k \geq l \geq 1$. The slab fits in the memory when $(N - g - h - t)k = lk^2 \leq M$; that is, $l \leq M/k^2$.

The following Table 2 gives the asymptotic I/O complexity of the l-th iteration, for $l \geq 1$. We omit the O-notation for brevity.

Let $c = \sqrt{M}/k$.

If $c \geq 1$, then the table simplifies as given in Table 3.

Summing over the iterations, the I/O complexity is as given in Table 2.2. For example, when $1 \leq \frac{(N-t)}{k} \leq c^2$, $\sum_{l=1}^{(N-t)/k} \alpha + \gamma = \sum_{l=1}^{(N-t)/k} Nlk/B = N(N - t)^2/Bk$.

Consider the following five statements: (The blank is to be filled in by the phrases listed below to obtain the five respective statements.) There exist values M, N and t with $t \leq \sqrt{M}$, (which makes it possible to choose $k = t$ with $c = \frac{\sqrt{M}}{k} = \frac{\sqrt{M}}{t} \geq 1$ satisfied) and ———— is less than the cost of $k = t$.

1. $t > \frac{M^2}{(N-t)^2}$ and so that the cost of $k = \frac{M}{(N-t)}$
2. $t > \frac{M^2}{(N-t)^2}$ and so that the cost of the best k in $(\frac{M}{N-t}, \frac{M^2}{(N-t)^2}]$

Table 2. α, γ and their sum in the l-th iteration for various values of l

l	α	γ	$\alpha + \gamma$
$1 \leq l \leq \frac{B}{k}$	k	$\frac{Nlk}{B}$	$\frac{Nlk}{B}$
$\frac{B}{k} < l \leq c^2$	$\frac{lk^2}{B}$	$\frac{Nlk}{B}$	$\frac{Nlk}{B}$
$c^2 < l \leq c^2\sqrt{k}$	$\frac{l^2 k^4}{BM}$	$\left(1 + \frac{1}{c}\right)\frac{Nlk}{B}$	$\left(1 + \frac{1}{c} + \frac{lk^3}{NM}\right)\frac{Nlk}{B}$
$c^2\sqrt{k} < l \leq c^2 k$	$\frac{lk^3}{B\sqrt{k}}$	$\left(1 + \frac{1}{c}\right)\frac{Nlk}{B}$	$\left(1 + \frac{1}{c} + \frac{k^2}{N\sqrt{k}}\right)\frac{Nlk}{B}$
$c^2 k < l$	$\frac{lk^3}{B}\left(\frac{1}{\sqrt{k}} + \frac{1}{\sqrt{M}}\right)$	$\left(1 + \frac{1}{c}\right)\frac{Nlk}{B}$	$\left(1 + \frac{1}{c} + \frac{k^2}{N\sqrt{k}} + \frac{k^2}{N\sqrt{M}}\right)\frac{Nlk}{B}$

Table 3. α, γ and their sum in the l-th iteration for various values of l, when $c \geq 1$

l	α	γ	$\alpha + \gamma$
$1 \leq l \leq \frac{B}{k}$	k	$\frac{Nlk}{B}$	$\frac{Nlk}{B}$
$\frac{B}{k} < l \leq c^2$	$\frac{lk^2}{B}$	$\frac{Nlk}{B}$	$\frac{Nlk}{B}$
$c^2 < l \leq c^2\sqrt{k}$	$\frac{l^2 k^4}{BM}$	$\frac{Nlk}{B}$	$\left(1 + \frac{lk^3}{NM}\right)\frac{Nlk}{B}$
$c^2\sqrt{k} < l$	$\frac{lk^3}{B\sqrt{k}}$	$\frac{Nlk}{B}$	$\left(1 + \frac{k^2}{N\sqrt{k}}\right)\frac{Nlk}{B}$

Table 4. The I/O complexity for Case 1 and $c \geq 1$, under various conditions

Condition	I/Os	Condition paraphrased
$1 \leq \frac{(N-t)}{k} \leq c^2$	$\frac{N(N-t)^2}{Bk}$	$k \leq \frac{M}{(N-t)}$
$c^2 < \frac{(N-t)}{k} \leq c^2\sqrt{k}$	$\frac{N(N-t)^2}{Bk} + \frac{k(N-t)^3}{BM} - \frac{M^2}{Bk^2}$	$\frac{M}{(N-t)} < k \leq \frac{M^2}{(N-t)^2}$
$c^2\sqrt{k} < \frac{(N-t)}{k}$	$\frac{N(N-t)^2}{Bk} + \frac{k(N-t)^2 - M^2}{B\sqrt{k}}$	$\frac{M^2}{(N-t)^2} < k$

3. $t > \frac{M^2}{(N-t)^2}$ and so that the cost of some k in $(\frac{M^2}{(N-t)^2}, t)$

4. $\frac{M}{N-t} < t \leq \frac{M^2}{(N-t)^2}$ and so that the cost of $k = \frac{M}{(N-t)}$

5. $\frac{M}{N-t} < t \leq \frac{M^2}{(N-t)^2}$ and so that the cost of some k in $(\frac{M}{N-t}, t)$

We claim that each of these statements is false. Proofs by contradiction follow:

Define $a = \log_M t$, $b = \log_M (N-t)$. Then $a, b > 0$, $t = M^a$, $(N-t) = M^b$ and $N = M^a + M^b$. Thus, $t \leq \sqrt{M} \Rightarrow a \leq 1/2$. For Statements 1, 2 and 3, therefore, $t > \frac{M^2}{(N-t)^2} \Rightarrow a > 2 - 2b \Rightarrow b > 1 - a/2 \Rightarrow b > 3/4$. So $N = (M^a + M^b) \approx M^b$.

For Statement 1, when $k = M/(N-t)$ the cost is $\frac{N(N-t)^2}{Bk} = NM^{3b-1}/B$.
When $k = t$, the cost is $\leq \frac{N(N-t)^2}{Bk} + \frac{k(N-t)^2}{B\sqrt{k}} = (NM^{2b-a} + M^{2b+a/2})/B$. But,
$NM^{3b-1} < M^{2b+a/2} \Rightarrow 4b - 1 < 2b + a/2 \Rightarrow (a/2 + 1) > 2b \geq 3/2 \Rightarrow a > 1$.
Contradiction.

The cost for k in $(\frac{M}{N-t}, \frac{M^2}{(N-t)^2}]$ is lower bounded by: $\frac{N(N-t)^2}{Bk} - \frac{M^2}{Bk^2}$ which is
minimised at $k = \frac{M^2}{(N-t)^2} = M^{2-2b}$. For Statement 2, the cost at $k = M^{2-2b}$ is
at most $(M^{5b-2} - M^{4b-2})/B \approx M^{5b-2}/B$. When $k = t$, the cost is, as before,
$(NM^{2b-a} + M^{2b+a/2})/B$. But, $M^{5b-2} < M^{2b+a/2} \Rightarrow 5b - 1 < 2b + a/2 \Rightarrow$
$(a/2 + 2) > 3b \geq 9/4 \Rightarrow a > 1/2$. Contradiction.

For Statement 3, the cost at t is $(NM^{2b-a} + M^{a/2+2b} - M^{2-a/2})/B$. The cost at
M^c, $\frac{M^2}{(N-t)^2} < M^c < t$ is $(NM^{2b-c} + M^{c/2+2b} - M^{2-c/2})/B$. But, $NM^{2b}(M^{-a} - M^{-c}) + M^{2b}(M^{a/2} - M^{c/2}) - M^2(M^{-a/2} - M^{-c/2}) > 0 \Rightarrow (2 - (a+c)/2 >$
$3b - a/2 - c) \vee (2b > 3b - a/2 - c) \Rightarrow b < 3/4$. Contradiction.

For statements 4 and 5, $M^{1-b} < M^a \leq M^{2-2b} \Rightarrow (1 - b) < a \leq 2 - 2b \Rightarrow b >$
$1/2$. That is, $a \leq 1/2 < b$.

Consider Statement 4. When $k = M/(N-t)$ the cost is $\frac{N(N-t)^2}{Bk} = NM^{3b-1}/B$.
When $k = t$, the cost is $\leq \frac{N(N-t)^2}{Bt} + \frac{t(N-t)^3}{BM} = (NM^{2b-a} + M^{a+3b-1})/B$. But,
$NM^{3b-1} < M^{a+3b-1} \Rightarrow N < M^a$. Contradiction.

For Statement 5, the cost at t is $(NM^{2b-a} + M^{a+3b-1} - M^{2-2a})/B$. The cost
at M^c, $\frac{M}{(N-t)} < M^c < t \leq \frac{M^2}{(N-t)^2}$, is $(NM^{2b-c} + M^{c+3b-1} - M^{2-2c})/B$. But,
$NM^{2b}(M^{-a} - M^{-c}) + M^{3b-1}(M^a - M^c) - M^2(M^{-2a} - M^{-2c}) > 0 \Rightarrow a + c >$
$1 \vee b < (2 - c)/3$. As $c < a \leq 1/2$, $a + c > 1$ cannot be. On the other hand,
$b < (2 - c)/3 \Rightarrow \frac{M}{(N-t)} = M^{(1-b)} > M^{(1+c)/3} \Rightarrow c > (1 + c)/3 \Rightarrow c > 1/2$;
contradiction.

What we have shown is that when $t \leq \sqrt{M}$, t is a better choice for k than
any of the smaller values. An analogous proof shows that if $\sqrt{M} < t$, then $\sqrt{M}$
is better than any smaller value.

If $c < 1$, (that is, $k > \sqrt{M}$) then Table 2.2 simplifies as given in the following
Table 2.2 and Table 6

Table 5. α, γ and their sum in the l-th iteration for various values of l, when $c < 1$

	α		γ		$\alpha + \gamma$	
$1 \leq l \leq c^2\sqrt{k}$	$\frac{l^2 k^4}{BM}$		$\frac{Nlk}{cB}$	$\left(\frac{1}{c} + \frac{lk^3}{NM}\right)\frac{Nlk}{B}$		
$c^2\sqrt{k} < l \leq c^2 k$	$\frac{lk^3}{B\sqrt{k}}$		$\frac{Nlk}{cB}$	$\left(\frac{1}{c} + \frac{k^2}{N\sqrt{k}}\right)\frac{Nlk}{B}$		
$c^2 k < l$	$\frac{lk^3}{B}\left(\frac{1}{\sqrt{k}} + \frac{1}{\sqrt{M}}\right)$		$\frac{Nlk}{cB}$	$\left(\frac{1}{c} + \frac{k^2}{N\sqrt{k}} + \frac{k^2}{N\sqrt{M}}\right)\frac{Nlk}{B}$		

Choosing $c < 1$ we would have $\sqrt{M} < k \leq t$. As the simplified Table 2.2 and
Table 6 show, any such choice would be worse than the choice of $k = \sqrt{M}$ in
the $c \geq 1$ case discussed above. That is, if $\sqrt{M} < t$, then $\sqrt{M}$ is better than any
larger value of $k \leq t$.

Table 6. The I/O complexity for Case 1 and $c < 1$, under various conditions

Condition	I/Os	Condition paraphrased
$1 \leq \frac{(N-t)}{k} \leq c^2\sqrt{k}$	$\frac{N(N-t)^2}{B\sqrt{M}} + \frac{k(N-t)^3}{BM}$	$k \leq \frac{M^2}{(N-t)^2}$
$c^2\sqrt{k} < \frac{(N-t)}{k} \leq c^2 k$	$\frac{N(N-t)^2}{B\sqrt{M}} + \frac{k(N-t)^2 - M^2}{B\sqrt{k}}$	$\frac{M^2}{(N-t)^2} < k$ and $(N-t) \leq M$
$c^2 k < \frac{(N-t)}{k}$	$\frac{N(N-t)^2}{B\sqrt{M}} + \frac{k(N-t)^2 - M^2}{B\sqrt{k}} + \frac{k(N-t)^2 - kM^2}{B\sqrt{M}}$	$\frac{M^2}{(N-t)^2} < k$ and $(N-t) > M$

If k is to be at most t, choosing it as $\min\{t, \sqrt{M}\}$ is the best.

Case 2: $k > t$ For $1 \leq i \leq (N-t)/k$ and $1 \leq j \leq k/t$, in the (i,j)-th iteration of the inner loop there are

- a QR decomposition of an $(N - g - h - t) \times t$ matrix,
- a matrix multiplication chain of dimensions $(h, N - g - t, t, t, t)$,
- a matrix multiplication chain of dimensions $(N-g-t, N-g-t, h+t, h+t, t)$, and
- a matrix multiplication chain of dimensions $(N-g-t, h+t, h+t, N-g-t, t)$,

where $g = (i-1)k$ and $h = (j-1)t$. Let α, β, γ and δ denote the I/O complexities of these operations, in that order. For $1 \leq i \leq (N-t)/k$, in the i-th iteration of the outer loop there is a

- a matrix multiplication chain of dimensions $(g+t, N-g-t, k, k, N-g-t)$,
- a matrix multiplication chain of dimensions $(N-g-t, N-g-t, k, k, N-g-k)$, and
- a matrix multiplication chain of dimensions $(N-g-t, k, k, N-g-t, N-g-k)$.

Let μ, ψ and ϕ denote the I/O complexities of these operations, in that order.

Let $c = \sqrt{M}/t$. Proceeding as in the analysis of case 1, we find that as i varies from 1 to $(N-t)/k$ and j varies from 1 to k/t, $(N-g-h-t)$ takes on values lt for $(N-t)/t \geq l \geq 1$, and therefore α in the l-th iteration is as given in the Table 7 below: (We omit the order notation for brevity.)

As $0 \leq g \leq N-t-k$, we have that $t \leq g+t \leq N-k$, $N-t \geq N-g-t \geq k$ and $N-k \geq N-g-k \geq t$. Therefore,

$$\sum_{i=1}^{(N-t)/k} \mu(i): O\left(\left(\frac{k}{\sqrt{M}} + 1\right)\sum_{i=1}^{(N-t)/k} \frac{(g+t)(N-g-t)}{B}\right),$$

$$\sum_{i=1}^{(N-t)/k} \psi(i): O\left(\left(\frac{k}{\sqrt{M}} + 1\right)\sum_{i=1}^{(N-t)/k} \frac{(N-g-t)^2}{B}\right), \text{ and}$$

$$\sum_{i=1}^{(N-t)/k} \phi(i): O\left(\left(\frac{k}{\sqrt{M}} + 1\right)\sum_{i=1}^{(N-t)/k} \frac{(N-g-k)(N-g-t)}{B}\right)$$

$$\sum_{i=1}^{(N-t)/k} \mu(i) + \psi(i) + \phi(i) \leq O\left(\left(\frac{k}{\sqrt{M}} + 1\right)\sum_{i=1}^{(N-t)/k} \frac{N(N-g-t)}{B}\right)$$

$$= O\left(\frac{N(N-t)^2}{kB}\left(\frac{k}{\sqrt{M}} + 1\right)\right)$$

Table 7. α in the l-th iteration for various values of l, when $k > t$

	α
$1 \leq l \leq \frac{B}{t}$	t
$\frac{B}{t} < l \leq c^2$	$\frac{lt^2}{B}$
$c^2 < l \leq c^2\sqrt{t}$	$\frac{l^2 t^4}{BM}$
$c^2\sqrt{t} < l \leq c^2 t$	$\frac{lt^3}{B\sqrt{t}}$
$c^2 t < l$	$\frac{lt^3}{B}\left(\frac{1}{\sqrt{t}} + \frac{1}{\sqrt{M}}\right)$

Similarly, $\sum_{i=1}^{(N-t)/k} \sum_{j=1}^{k/t} \gamma(i,j) = O\left(\sum_{i=1}^{(N-t)/k} \sum_{j=1}^{k/t} \frac{(N-g-t)^2}{B}\left(\frac{t}{\sqrt{M}}+1\right)\right)$.

Also, $\sum_{i=1}^{(N-t)/k} \sum_{j=1}^{k/t} \beta(i,j) + \delta(i,j) = O\left(\sum_{i=1}^{(N-t)/k} \sum_{j=1}^{k/t} \frac{(N-g-t)(h+t)}{B}\left(\frac{t}{\sqrt{M}}+1\right)\right)$

As $N - g - t \geq h + t$, we have:

$$\sum_{i=1}^{(N-t)/k} \sum_{j=1}^{k/t} \beta(i,j) + \gamma(i,j) + \delta(i,j) \leq O\left(\left(\frac{t}{\sqrt{M}}+1\right) \sum_{i=1}^{(N-t)/k} \sum_{j=1}^{k/t} \frac{(N-g-t)^2}{B}\right)$$

$$= O\left(\left(\frac{t}{\sqrt{M}}+1\right) \frac{k}{t} \sum_{i=1}^{(N-t)/k} \frac{(N-g-t)^2}{B}\right) = O\left(\left(\frac{(N-t)^3}{tB}\left(\frac{t}{\sqrt{M}}+1\right)\right)\right)$$

The total cost of all matrix multiplications is: $O\left(\frac{N(N-t)^2}{B}\left(\frac{1}{\sqrt{M}} + \frac{1}{t} + \frac{1}{k}\right)\right)$.

The total cost of QR-decomposition is:

As can be seen, the I/O complexity, the sum of the above two, is independent of the choice of k. But, when $t \leq \sqrt{M}$, a choice of $k = t$, and when $t > \sqrt{M}$, a choice of $k = \sqrt{M}$ would give lower costs than these.

Thus, we conclude that the slab based algorithm does the best when k is chosen as $\min\{\sqrt{M}, t\}$.

Table 8. The total cost of QR-decomposition, under various conditions

Condition	Cost of QR decomposition
$1 \leq \frac{(N-t)}{t} \leq \frac{B}{t}$	$N - t$
$\frac{B}{t} < \frac{(N-t)}{t} \leq c^2$	$\frac{(N-t)^2}{B} + B$
$c^2 < \frac{(N-t)}{t} \leq c^2\sqrt{t}$	$\frac{t(N-t)^3}{BM}$
$c^2\sqrt{t} < \frac{(N-t)}{t} \leq c^2 t$	$\frac{t(N-t)^2 - M^2}{B\sqrt{t}}$
$c^2 t < \frac{(N-t)}{t}$	$\frac{t(N-t)^2 - M^2}{B\sqrt{t}} + \frac{t(N-t)^2 - tM^2}{B\sqrt{M}}$

3   Conclusion

Both the unblocked and a blocked direct Hessenberg reduction algorithms take $O(N^3)$ flops and $O(N^3/B)$ I/Os. For large matrices, the performance can be improved on machines with multiple levels of memory, by the two step reduction, since all most all the operations of the first step are matrix-matrix operations. We show that reduction of a nonsymmetric matrix to banded Hessenberg form of bandwidth t takes $O(N^3/\min\{t, \sqrt{M}\}B)$ I/Os. We also show that the slab based algorithm does the best when the slab width k is chosen as $\min\{\sqrt{M}, t\}$. It is also observed that, in the existing slab based algorithms, some of the elementary matrix operations like matrix multiplication should be handled I/O efficiently, to achieve optimal I/O performances.

References

1. Aggarwal, A., Vitter, J.S.: The input/output complexity of sorting and related problems. Comm. ACM 31(9), 1116–1127 (1988)
2. Vitter, J.S.: External memory algorithms. In: Handbook of Massive Data Sets. Massive Comput., vol. 4, pp. 359–416. Kluwer Acad. Publ., Dordrecht (2002)
3. Mohanty, S.K.: I/O Efficient Algorithms for Matrix Computations. PhD thesis, Indian Institute of Technology Guwahati, Guwahati, India (2010)
4. Mohanty, S.K., Sajith, G.: I/O efficient QR and QZ algorithms. In: 19th IEEE Annual International Conference on High Performance Computing (HiPC 2012), Pune, India (accepted, December 2012)
5. Roh, K., Crochemore, M., Iliopoulos, C.S., Park, K.: External memory algorithms for string problems. Fund. Inform. 84(1), 17–32 (2008)
6. Chiang, Y.J., Goodrich, M.T., Grove, E.F., Tamassia, R., Vengroff, D.E., Vitter, J.S.: External-memory graph algorithms. In: Proceedings of the Sixth Annual ACM-SIAM Symposium on Discrete Algorithms, pp. 139–149. ACM, Philadelphia (1995)
7. Chiang, Y.J.: Dynamic and I/O-Efficient Algorithms for Computational Geometry and Graph Problems: Theoretical and Experimental Results. PhD thesis, Brown University, Providence, RI, USA (1996)
8. Goodrich, M.T., Tsay, J.J., Vengroff, D.E., Vitter, J.S.: External-memory computational geometry. In: Proceedings of the 34th Annual IEEE Symposium on Foundations of Computer Science, pp. 714–723. IEEE Computer Society Press, Palo Alto (1993)
9. Arge, L.: The buffer tree: a technique for designing batched external data structures. Algorithmica 37(1), 1–24 (2003)
10. Vitter, J.S.: External memory algorithms and data structures: dealing with massive data. ACM Comput. Surv. 33(2), 209–271 (2001)
11. Demaine, E.D.: Cache-oblivious algorithms and data structures. Lecture Notes from the EEF Summer School on Massive Data Sets, BRICS, University of Aarhus, Denmark (2002)
12. Vitter, J.S., Shriver, E.A.M.: Algorithms for parallel memory. I. Two-level memories. Algorithmica 12(2-3), 110–147 (1994)
13. Toledo, S., Gustavson, F.G.: The design and implementation of SOLAR, a portable library for scalable out-of-core linear algebra computations. In: Fourth Workshop on Input/Output in Parallel and Distributed Systems, pp. 28–40. ACM Press (1996)

14. Reiley, W.C., Van de Geijn, R.A.: POOCLAPACK: parallel out-of-core linear algebra package. Technical Report CS-TR-99-33, Department of Computer Science, The University of Texas at Austin (November 1999)
15. Alpatov, P., Baker, G., Edwards, H.C., Gunnels, J., Morrow, G., Overfelt, J., de Geijn, R.A.V.: PLAPACK: Parallel linear algebra package design overview. In: Supercomputing 1997: Proceedings of the ACM/IEEE Conference on Supercomputing, pp. 1–16. ACM, New York (1997)
16. Van de Geijn, R.A., Alpatou, P., Baker, G., Edwards, C., Gunnels, J., Morrow, G., Overfelt, J.: Using PLAPACK: Parallel Linear Algebra Package. MIT Press, Cambridge (1997)
17. Choi, J., Dongarra, J.J., Pozo, R., Walker, D.W.: ScaLAPACK: A scalable linear algebra library for distributed memory concurrent computers. In: Proceedings of the Fourth Symposium on the Frontiers of Massively Parallel Computation, pp. 120–127. IEEE Computer Society Press (1992)
18. Anderson, E., Bai, Z., Bischof, C.H., Demmel, J., Dongarra, J.J., Croz, J.D., Greenbaum, A., Hammarling, S., McKenney, A., Ostrouchov, S., Sorensen, D.C.: LAPACK Users' Guide, 2nd edn. SIAM, Philadelphia (1995)
19. Basic Linear Algebra Subprograms(BLAS), http://www.netlib.org/blas/
20. Toledo, S.: A survey of out-of-core algorithms in numerical linear algebra. In: External Memory Algorithms. DIMACS Ser. Discrete Math. Theoret. Comput. Sci. Amer. Math. Soc., vol. 50, pp. 161–179, Piscataway, NJ, Providence, RI (1999)
21. Elmroth, E., Gustavson, F.G., Jonsson, I., Kågström, B.: Recursive blocked algorithms and hybrid data structures for dense matrix library software. SIAM Rev. 46(1), 3–45 (2004)
22. Haveliwala, T., Kamvar, S.D.: The second eigenvalue of the google matrix. Technical Report 2003-20, Stanford InfoLab (2003)
23. Christopher, M.D., Eugenia, K., Takemasa, M.: Estimating and correcting global weather model error. Monthly Weather Review 135(2), 281–299 (2007)
24. Alter, O., Brown, P.O., Botstein, D.: Processing and modeling genome-wide expression data using singular value decomposition. In: Bittner, M.L., Chen, Y., Dorsel, A.N., Dougherty, E.R. (eds.) Microarrays: Optical Technologies and Informatics, vol. 4266, pp. 171–186. SPIE (2001)
25. Xu, S., Bai, Z., Yang, Q., Kwak, K.S.: Singular value decomposition-based algorithm for IEEE 802.11a interference suppression in DS-UWB systems. IEICE Trans. Fundam. Electron. Commun. Comput. Sci. E89-A(7), 1913–1918 (2006)
26. Golub, G.H., Van Loan, C.F.: Matrix Computations, 3rd edn. Johns Hopkins Studies in the Mathematical Sciences. Johns Hopkins University Press, Baltimore (1996)
27. Watkins, D.S.: Fundamentals of Matrix Computations, 2nd edn. Pure and Applied Mathematics. Wiley-Interscience. John Wiley & Sons, New York (2002)
28. Dongarra, J.J., Duff, I.S., Sorensen, D.C., Van der Vorst, H.A.: Numerical Linear Algebra for High Performance Computers. Software, Environments and Tools, vol. 7. SIAM, Philadelphia (1998)
29. Dongarra, J.J., Croz, J.D., Hammarling, S., Duff, I.S.: A set of level 3 basic linear algebra subprograms. ACM Trans. Math. Softw. 16(1), 1–17 (1990)
30. Elmroth, E., Gustavson, F.G.: New Serial and Parallel Recursive QR Factorization Algorithms for SMP Systems. In: Kågström, B., Elmroth, E., Waśniewski, J., Dongarra, J. (eds.) PARA 1998. LNCS, vol. 1541, pp. 120–128. Springer, Heidelberg (1998)
31. Gunter, B.C., Reiley, W.C., Van de Geijn, R.A.: Implementation of out-of-core Cholesky and QR factorizations with POOCLAPACK. Technical Report CS-TR-00-21, Austin, TX, USA (2000)

32. Gunter, B.C., Reiley, W.C., Van De Geijn, R.A.: Parallel out-of-core Cholesky and QR factorization with POOCLAPACK. In: IPDPS 2001: Proceedings of the 15th International Parallel & Distributed Processing Symposium. IEEE Computer Society, Washington, DC (2001)

33. Gunter, B.C., Van de Geijn, R.A.: Parallel out-of-core computation and updating of the QR factorization. ACM Trans. Math. Software 31(1), 60–78 (2005)

34. Buttari, A., Langou, J., Kurzak, J., Dongarra, J.J.: A class of parallel tiled linear algebra algorithms for multicore architectures. Parallel Comput. 35(1), 38–53 (2009)

35. Bischof, C.H., Lang, B., Sun, X.: A framework for symmetric band reduction. ACM Trans. Math. Software 26(4), 581–601 (2000)

36. Quintana Ortí, G., de Geijn, R.A.V.: Improving the performance of reduction to Hessenberg form. ACM Trans. Math. Software 32(2), 180–194 (2006)

37. Dongarra, J.J., Sorensen, D.C., Hammarling, S.J.: Block reduction of matrices to condensed forms for eigenvalue computations. J. Comput. Appl. Math. 27(1-2), 215–227 (1989)

38. Dongarra, J.J., van de Geijn, R.A.: Reduction to condensed form for the eigenvalue problem on distributed memory architectures. Parallel Comput. 18(9), 973–982 (1992)

39. Bischof, C.H., Lang, B., Sun, X.: Parellel tridiagonal through two-step band reduction. In: Proceedings of the Scalable High-Performance Computing Conference, pp. 23–27. IEEE Computer Society Press (May 1994)

40. Lang, B.: Using level 3 BLAS in rotation-based algorithms. SIAM J. Sci. Comput. 19(2), 626–634 (1998)

41. Lang, B.: A parallel algorithm for reducing symmetric banded matrices to tridiagonal form. SIAM J. Sci. Comput. 14(6), 1320–1338 (1993)

42. Berry, M.W., Dongarra, J.J., Kim, Y.: A parallel algorithm for the reduction of a nonsymmetric matrix to block upper-Hessenberg form. Parallel Comput. 21(8), 1189–1211 (1995)

43. Ltaief, H., Kurzak, J., Dongarra, J.J.: Parallel block Hessenberg reduction using algorithms-by-tiles for multicore architectures revisited. LAPACK Working Note #208, University of Tennessee, Knoxville (2008)

44. Bai, Y., Ward, R.C.: Parallel block tridiagonalization of real symmetric matrices. J. Parallel Distrib. Comput. 68(5), 703–715 (2008)

45. Großer, B., Lang, B.: Efficient parallel reduction to bidiagonal form. Parallel Comput. 25(8), 969–986 (1999)

46. Lang, B.: Parallel reduction of banded matrices to bidiagonal form. Parallel Comput. 22(1), 1–18 (1996)

47. Trefethen, L.N., Bau III, D.: Numerical Linear Algebra. SIAM (1997)

48. Ltaief, H., Kurzak, J., Dongarra, J.J.: Scheduling two-sided transformations using algorithms-by-tiles on multicore architectures. LAPACK Working Note #214, University of Tennessee, Knoxville (2009)

49. Bischof, C.H., Van Loan, C.F.: The WY representation for products of Householder matrices. SIAM J. Sci. Statist. Comput. 8(1), S2–S13 (1987)

50. Wu, Y.J.J., Alpatov, P., Bischof, C.H., van de Geijn, R.A.: A parallel implementation of symmetric band reduction using PLAPACK. In: Proceedings of Scalable Parallel Library Conference. PRISM Working Note 35, Mississippi State University (1996)

51. Bai, Y.: High performance parallel approximate eigensolver for real symmetric matrices. PhD thesis, University of Tennessee, Knoxville (2005)

Luring Conditions and Their Proof of Necessity through Mathematical Modelling

Anand Gupta, Prashant Khurana, and Raveena Mathur

Division of Information Technology
Netaji Subhas Institute of Technology, University of Delhi, New Delhi, India
{omaranand,prashantkhurana145,raveenamathur91}@gmail.com

Abstract. Luring is a social engineering technique used to capture individuals having malicious intent of breaching the information security defense of an organization. Certain conditions(Need, Environment, Masquerading Capability and Unawareness) are necessary for its effective implementation. To the best of our knowledge the necessity of these conditions is not yet proved so far. The proof is essential as it not only facilitates automation of the luring mechanism but also paves way for proof of the completeness of the conditions. The present paper attempts on this aspect by invoking three approaches namely probability, entropy and proof by contra positive. Also, the concept of cost effectiveness is introduced. Luring is acceptable if its cost works out less than cost of data theft.

Keywords: Social Engineering, Luring, Honeypot, Contrapositive, Entropy, Bayesian Graphs, Probability.

1 Introduction

The greatest security threat that organizations are facing is the connivance of social engineering attacks to breach information security defenses of an organization. Possible methods include obtaining the password of an authentic user, gaining his trust and infringing his own rights. According to statistics in [1] 48 % of large companies and 32 % of the companies of all sizes have experienced 25 or more such attacks in the past two years.

For defending against such attacks, [2] proposes a solution based on the age old social engineering technique called Luring, a practice of enticing the suspect into context honeypot. However, creation of a luring environment poses implementation difficulties as it has to be conducive and credible. It requires meticulous understanding about what it is that makes a suspect get attracted to a context honeypot or what are the essential conditions to ensure effective luring. These conditions (viz. Need, Masquerading Capability, Environment, Unawareness about Target Data) are explored by Gupta et al[2].

This paper addresses the question whether these conditions are necessary. For this, we have defined each of the conditions and carried out the following:

S. Srinivasa and V. Bhatnagar (Eds.): BDA 2012, LNCS 7678, pp. 148–157, 2012.

- Qualitative analysis: Each condition is qualitatively analysed by doing social and behavioural research on the suspect. The proof of necessity is furnished through a logical analysis of the impact on luring in the absence of each of these conditions one by one. This is presented as proof by contra positive.
- Quantitative analysis: A luring environment if predictable is ineffective. Hence quantification of predictability using entropy is followed as a method of proof.

Beside the aforesaid introductory remarks, the paper is organized as follows: Section 2 describes the related work pertaining to our research work. Section 3 elucidates the proof by contra positive. Section 4 gives proof of necessity using Bayesian probability and its subsequent analysis using Shannons entropy. Section 5 concludes the paper, while Section 6 describes the future work.

2   Related Work and Motivation

2.1   Necessary Conditions

Several approaches both statistical and knowledge based have been proposed for identification of essential conditions. Their identification makes the system deterministic and implementable, thus facilitating the analysis of system behaviour.

In context of luring, similar work is presented by Gupta et al[2]. Herein, the essential conditions are identified and based on them an architecture for generating lure messages is developed. However, simple observations from the architecture cannot help in standardization of the luring system. So, the need for mathematically modelling the conditions is felt by us. Modelling paves the way for the proof of necessity and depicts the interdependencies. It further forms the basis for security threat analysis.

2.2   Proof by Contra Positive

Any study on proof by Contra Positive establishes the truth or validity of a proposition by demonstrating the same of the converse of its negated parts[3]. In [4], the authors have used this method to show that Bergstrom's result does not get generalized to the benefit-cost analysis of generic changes in public goods and that there may exist good projects to be rejected by a selfish-benefit cost test. Other studies involving the principle of contra positive have been carried out in [5] where the authors have investigated proof theory of default reasoning and have proved deduction theorems for default iconic logic.

2.3   Entropy

There have been several approaches on application of entropy to determine the information content of a system. In a study of entropy by Shannon [6], predictability of a piece of text is used to eliminate redundancy for transmission. It along with other experimental results is used to formulate Shannon's entropy and its interpretation of predictability.

The papers by Shannon[6] and Maasoumi, et al[7] have inspired us, to make use of predictability theory to find out whether the luring environment would lure or not. The system is considered to fail whenever it becomes predictable to the suspect. The predictability is then mathematically calculated using Shannons entropy to be further interpreted in Section 4 of this paper.

3   Proof Using Contra positive

We start by looking at the conditions for luring as proposed by [2] and define them formally and mathematically.

3.1   Definitions of Condition

1. Masquerading capability (M):It is defined as the ability of a person having a given malafide intention and possessing a username and a password to fool the system that he is an authentic user. If X is the finite set of all authorized users x_i , then $x_i \in X$, and if there is a person A with a Malafide Intention, then the ease with which $A \equiv x_i$ (A pretends to be x_i) can be proven to the system, is defined as his masquerading capability.

2. Unawareness about target data (U): It can be defined as the extent of knowledge possessed by or accessible to the intruder. If in the honeypot H, x_i wants to access a set of data D such that $D \subseteq H$ and if he already possesses data R, then he is said to be totally unaware of the data D iff $D \cap R = \emptyset$. Further, we can quantitatively define

$$U = 1 - \frac{(D \cap R)}{(D)} \tag{1}$$

3. Local Environment (L):It can be defined as the environment which can be directly influenced by H and its defense and luring systems. Formally it can be said that, if 'n' is the total number of elements directly in control of H, and n_i is the total number of elements which are turned into lures for the intruder, then

$$L = \frac{n_i}{n} \tag{2}$$

4. Global Environment (G):We introduce this characteristic as a replacement for both Need and Environment as defined in [2]. G is formally defined as −H such that $H \cup G = U$ and $H \cap G = \emptyset$. Clearly both Need and Environment as defined in the paper are contained in the set G.

5. Data Value (DV):It is a new concept as introduced by us. It keeps track of the value of each data for the creator of H. Formally, DV is defined for each data H_i as the direct financial and social loss faced by the creator of H in case H_i is obtained by the intruder.

6. Other Conditions (OC):These are the conditions which are yet undiscovered.

3.2 Proving Necessity of the Conditions

To prove the necessity, we stipulate two assumptions, which are:

- The intruder must be rational, i.e., he must not intend to get caught and must not risk his own safety unless the utility of the information sought is far greater than any possible adversity.
- Luring must be efficient, i.e., cost of luring must be less than the cost of the protecting the data.

1. Proof of Requirement for M:Consider the set C of conditions. Let us say that in the given scenario, the set C can be defined as $\{\neg M, U, G, L, DV, OC\}$ where OC may or may not be $\emptyset$. Since $\neg M$ implies that the system does not offer any masquerading capability, so it leads to two possibilities.

   Going back to the definition, either $\neg \exists x_i$, $x_i \notin X$, which implies that the entire information is publicly available and hence no need for masquerading and intrusion arise. Evidently, luring is not possible here, as it in itself violates another condition U.

   The other possibility is that $\neg \exists x_i$, $x_i \in X$. In fact the intrusion detection system is so powerful that it does not allow any masquerading to occur. Any intrusion in such a situation directly violates our axiom since the intruder knows that any act of masquerading will be detected before he could obtain any information. Thus it renders any potential benefit to zero and opens him to a possible attack. Since his own safety is a prime consideration, so he will not be lured.

2. Proof of Requirement for U:Again consider C, which in this case would be $\{M, \neg U, G, L, DV, OC\}$. Going back to the standard definition proposed, we can say that in the case of $\neg U, D \cap R = R$ where R is the knowledge the intruder already holds. Such a scenario directly conflicts with the assumption. It is because any attempt of intrusion will put him at some small risk δ while he will not really gain anything. Hence he will not enter the system or would not be lured.

3. Proof of Requirement for L:C, in this case, would be $\{M, UL, G, \neg L, DV, OC\}$. The lack of a local environment would mean that the context honeypot is isolated and that no elements that have a direct linkage with H. Clearly, in such a case luring is trivially impossible since there is no place for the intruder to be lured to.

4. Proof of Requirement for G:The proof in this case, consists of two parts. The first part shows that the condition 'need' discussed in [2] is in fact, a proper subset of G. Subsequently, it is proven that G is a necessary condition for efficient luring.

   To prove that Need, $N \subset G$, consider any individual x_i such that $x_i \in X$ where X is the set of people authorized to access the confidential data, D. In case any intruder has a need N to access a part of D, clearly N as defined for him in [2] is non zero. However, consider the variable G, which would be calculated for the entire set X and the data D and not just the individual x_i. Clearly, when n = 1, G reduces to N.

Further, when G is not present, i.e. when C is of the form $\{M, U, \neg G, L, DV, OC\}$, the situation is that there are no intruders to be trapped for the context honeypot. Clearly if there is no one to be lured, the basic premise of the luring system itself fails.

5. Proof of Requirement for DV:The cost involved in luring a potential intruder is high and involves usage of both processing power as well as space. The condition of Data Value trivially ensures efficient luring by simply calculating the potential loss considered after examining the existing knowledge possessed by the intruder and the value of the data that the intruder is trying to obtain.

4   Entropy Model

This section furnishes another view of the Luring system, wherein a deductive analysis using mathematical modelling suffices for a proof of necessity. For this, approach the luring conditions are redefined and bifurcated, as a new view of the luring system is generated keeping in mind the mathematical coherence.

4.1   Environment

It is the framework of any luring model. We create an environment to lure the person to jump into the system and not recede during the process. For this we create certain conditions which tempt the user into entering and completing the process. Hence, we create lures for the user and the environment E is defined as in (2).These factors might enforce need or masquerading capability or the awareness of the user. We consider environment to be of three types:

1. **Masquerading Environment(ME):** It enforces the masquerading capability of the user. ME is defined as:

$$ME = \frac{\text{Number of masquerading factors turned into lure}}{\text{Total number of masquerading factors}} \tag{3}$$

2. **Target Data Environment(TDE):** It enforces the awareness of the user. It is defined in similar fashion as Masquerading Environment:

$$TDE = \frac{\text{Number of target data factors turned into lure}}{\text{Total number of target data factors}} \tag{4}$$

3. **Need Enforcement Environment(NE):** It enforces need of the person, and is defined in the similar fashion as:

$$NE = \frac{\text{Number of need factors turned into lure}}{\text{Total number of need factors}} \tag{5}$$

The impact of environment upon the success or failure of a luring model will be a function which initially increases and then decreases after the mid point(eg. Sinusoidal function in domain 0 to π).Also the function must be in range 0 to

1. To justify this, we see that with the increase in the number of factors turned into lure, the probability of success increase initially. Further, as more factors are turned into lures, more users get tempted into the model, indicating more are the chances of success of our model. But as more factors are turned into lures, the predictability also keeps on increasing, because so many baits may create a doubt in the mind of the user regarding the authenticity of the model. He may begin to feel as if he is being trapped and hence might not enter the system or might opt out in the middle.

4.2 Need

Need is both inherent in a user and is also enforced by the environment (Need environment). Hence total need is the sum of enforced need and the inherent need.

$N = x + NE$, where x is the inherent need and NE is the value of Need environment which enforces need.

The impact of need upon success or failure of the luring model will be a function similar in properties as that of environment but should increase as N goes from 0 to 1(eg.a Sinusoidal function in domain 0 to $\pi/2$). To justify it we see that more is the need of a user, more are the chances of him getting into the model and completing the required steps. Let this function be f.

$$P(L/N) = f \tag{6}$$

4.3 Masquerading Capability

It is both inherent in a user and is also enforced by the environment (Masquerading capability environment).We define the following terms

1. Masquerading capability Inherent(MCI):It is initially present with the person under test.
2. Masquerading Capability Enforced(MCE):It is enforced by the Environment. Let the value of Masquerading capability environment be x, then MCE is a linearly increasing function of x.
3. Masquerading capability Present(MCP):It is the sum of inherent and enforced masquerading capability i.e. MCP = MCI + MCE
4. Masquerading capability Required(MCR):It is the capability to enter the honeypot and remain inside it until all the steps are complete.
5. Masquerading Capability to be Gained(MCG):It represents the difference between the masquerading capability required and inherent masquearding capability of the user i.e. MCG is the MCG = MCR − MCP.

The impact of masquerading capability depends upon the MCG.

- MCG = 0, means that the user will complete all steps of the model without any difficulty.
- MCG > 0, means that the user will have to gain the required MCG. More is the MCG, more are the chances of the user dropping in the middle of the process as he doesn't have the required capability.

- MCG < 0, means that the user has additional masquerading capability, even though he might complete all steps of the model, it would be difficult to catch him. Higher is the negative value, more are the chances of user escaping unnoticed.

Impact of masquerading capability is similar to that of environment(eg. Sinusoidal function in domain 0 to π). Let this function be g.

$$P(L/M) = g \tag{7}$$

4.4 Unawareness about Target Data

Awareness about target data is inherent in a user and is enforced by the environment(Target Data Environment). Impact of unawareness is similar to that of environment(eg.a sinusoidal function from 0 to π).

Let α be the inherent value of awareness in the user, then total awareness $TDA = \text{TDAE} + \alpha$ and Unawareness (U) $= 1 - \text{Awareness} = 1\text{-TDA}$. More is the unawareness, better will be the system as user is less conspicuous.

4.5 Data Value

It signifies value of data which can be protected while ensuring the success of the model. More the unawareness of the user, more the data value.

Let the worth of the data be D.Then data value $DV = D * U$, as this is the amount of data can be protected.

The impact of data value is similar to that of need(eg.a sinusoidal function from 0 to $\pi/2$). Let this function be h.

$$P(L/DV) = h \tag{8}$$

4.6 Efficiency of Luring Model

Efficiency of a luring model is defined as the ability of a given luring model to lure a given suspected user. Using the formula's in 4.2,4.3 and 4.5.

$$P(L) = P(L/N) * P(L/M) * P(L/DV)$$
$$= f * g * h$$

Environment is not taken into account as it directly effects all of the other conditions which are used in the formula. Also unawareness is not taken into account as data value is used which encompasses unawareness.

4.7 Proving Necessity

A luring environment lures a person. If it is predictable to the suspect/user, in terms of the information provided by it, then it fails to lure. By predictability, it is meant that the suspect can predict the outcome of the luring mechanism i.e.

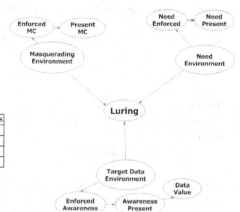

Table 1. States and Values

States	Masquerading Capability	Need	Unawareness
State1	3	3	1
State2	4	4	1
State3	5	5	1
State4	6	2	1

Fig. 1. Bayesian Graph

at any state he may predict he must leave it(useless to him),or he may predict that he is getting lured. But if he can't predict the outcome then he chooses to remain in the system, to be guided by the luring mechanism designer. It is for us to discover, based on the suspect and his retention period that whether he will get lured or not. For this purpose, Shannon's entropy a measure of information content of the system, is used. The entropy so found is analyzed to conclude about the predictability of the luring environment, given the suspect. The analysis applies the notion that a system is said to be predictable if its entropy value is lower than a benchmark value.

4.7.1   Luring Environment/System States
System is defined into states, each state having a specific value of masquerading capability, need and unawareness required to reach that state. Suspect goes from State 1→State 2→State 3→State 4. Table 1 depict the state values. The values in the table are calculated using the formulae discussed in section 4.1.

4.7.2   Proving Necessity
Shannon's entropy is defined as

$$h(x) = -p(x) \log p(x)$$

1. Entropy of the system:The entropy of the system is defined as h(L).
   h(L)=h(MC)+h(N)+h(U), according to the Bayesian network(fig. 1)
   where, $h(MC)$= -$p(MC)log$ $p(MC)$, $h(N)$= -$p(N)log$ $p(N)$, $h(U)$= -$p(U)log$
   $p(U)$(Entropies are added as MC, N and U are independent).
      The Benchmark value is chosen by considering that after reaching State 3(for this example)there is a very high probability that the suspect will be lured. The state 3 in this example(or any other state) is carefully designed by the luring mechanism designer to have the highest entropy value. It is

done so as to ensure that the suspect with same parameters as state 3 or $h(1)=9.3$,can not predict the outcome of the process he is currently enrolled for. (A result of Shannon's Entropy and predictability implication, according to which the maximum entropy state is the one with minimum predictability)

$h(L)= - [(0.75log(0.75))+(0.75log0.75)+(0.75log0.75)]$

$h(L)= 9.33$.

Hence this is the bench mark value.

2. Masquerading Environment:The entropy due to Masquerading Environment is given by:

$h(MC) = -p(MC) log p(MC)$

We know,

MC present = inherent MC + enforced MC

• Case 1: MC is low , typically less to reach even state 3.Entropy $(h(L))$ of the system decreases, as $h(MC)$ becomes low. System is predictable to him.We apply the same to the example of table 1:

If, suppose the enforced masquerading capability is 2 then $p(MC=3)=1/ 4$. Now, need and unawareness are so chosen such that the person is just able to cross the state 3. Hence, net need of the person is 4 (inclusive of inherent and enforced). Similarly unawareness is chosen to be 1.

$h(L)=0.5+3.075+3.075=6.65$

$h(L)<$ benchmark value.

• Case 2: Masquerading capability is high. For the example in Table 1, if the masquerading capability is 9,

then p(MC=9)=1 and h(luring)=0+3.075+.075=6.15

We find that h(luring)< benchmark value. Hence, the suspect with a very high masquerading capability will not be lured.He can predict the outcome of the system , that is, he is getting lured.

3. Need Based environment:The entropy due to need is given by:

h(N)= -p(N)log p(N)

Suppose the need is low, for the example of table 1.

If inherent need =2, enforced need is 1, then net need present is 3,

Then $h(L)= -[(0.25)log(1/4)+.75log(0.5)+.75log(0.75)]$

$h(L)=6.65<$ benchmark value. Hence, the system becomes predictable to the suspect and he predicts he is being lured. Thus, absence or very rare presence of need cannot lure. An argument to the case when net need value is very high follows a similar proof (case 2 of Masquerading capability)

4. Target data based Environment:We can similarly show that as unawareness about target data decreases, predictability increases. It is attributed to the fact that a highly aware suspect will easily detect the presence of synthetic data that is used in baits. Thus the absence of unawareness about target data will not materialize luring of the suspect.

We may not consider the case when unawareness of the suspect is very high as it is not possible as during the course of the luring mechanism it is only decreased(luring environment enforces awareness) and one's inherent awareness can't be extremely low, it would not go in conformity with his aim of data theft, as he will always need to have some information of his interest.

We can thus conclude that, if any of these three parameters is absent in the suspect or present in sparse quantities, then the suspect will not be lured.

5   Conclusion

In this paper we have mathematically defined the conditions required for luring and proved their necessity. The mathematical model has enabled us to determine the state suspect will go till he becomes suspicious of the system. We are able to answer why a system has failed and propose the improvements required for luring to be more robust and effective. In addition to proving the necessity of existing conditions we have kept in mind the efficiency of the luring system i.e cost of luring is less than the cost of data theft.

6   Future Work

The mathematical models so presented in the paper are built on certain assumptions. We intend to optimize our models by minimizing these assumptions. Further, a formalized polynomial should exist which can provide a reference base to calculate the efficiency of luring in a given scenario. New conditions which are not the subset of existing conditions could be explored and experiments based on the model can be conducted to collect the empirical data.

References

1. Check Point Software Technologies Survey on, The Risk of Social Engineering on Information Security: A survey of IT Professional Dimensional Research (September 2011), http://www.checkpoint.com/surveys/socialeng1509/socialeng.htm
2. Gupta, S.K., Gupta, A., Damor, R., Goyal, V.: Luring: A framework to induce a suspected user into Context Honeypot. In: Proceedings of the 2nd International Workshop on Digital Forensics and Incident Analysis (WDFIA 2007), Samos, Greece, August 27-28, pp. 55–64. IEEE Computer Society (2007)
3. Wikipedia; Information about Proof by Contrapositive, http://en.wikipedia.org/wiki/Proof_by_contrapositive
4. Flores, N.E.: Non-Paternalistic Altruism and Welfare Economics. Journal of Public Economics 83(2), 293–305 (2002)
5. Nait Abdalla, M.A.: An Extended Framework for Default Reasoning. In: Csirik, J.A., Demetrovics, J. (eds.) FCT 1989. LNCS, vol. 380, pp. 339–348. Springer, Heidelberg (1989)
6. Shannon, C.E.: Prediction and entropy of printed English. Bell Systems Technical Journal 30, 50–64 (1951)
7. Maasoumi, E., Racine, J.: Entropy and predictability of stock market returns. Journal of Econometrics 107(1-2), 291–312 (2002)

Efficient Recommendation for Smart TV Contents

Myung-Won Kim, Eun-Ju Kim, Won-Moon Song, Sung-Yeol Song, and A. Ra Khil

Department of Computer Science and Engineering,
Soongsil University, Seoul, Korea
{mkim,blue7786,gtangel,randol,ara}@ssu.ac.kr

Abstract. In this paper, we propose an efficient recommendation technique for smart TV contents. Our method solves the scalability and sparsity problems from which the conventional algorithms suffer in smart TV environment characterized by the large numbers of users and contents. Our method clusters users into user groups of similar preference patterns and a set of similar users to the target user are extracted, and then the user-based collaborative filtering is applied. We experimented with our method using the data of the real one-month IPTV services. The experiment results showed the success rate of 93.6% and the precision of 77.4%, which are recognized as a good performance for smart TV. We also investigate integration of recommendation methods for more personalized and efficient recommendation. Category match ratios for different integrations are compared as a measure for personalized recommendation.

Keywords: recommendation, collaborative filtering, clustering, smart TV, data mining, ISOData.

1   Introduction

With the recent development of communication and broadcasting technologies smart TV emerges and it provides a wide variety of services including high-quality telecasting contents, two-way communication, information retrieval, Internet shopping, and Internet games [1]. Smart TV not only telecasts the existing air TV or cable TV, but also provides contents such as domestic and foreign movies, drama series, entertainment, and video-on-demand contents. Consequently, users can freely choose contents that they prefer, but they may experience difficulties to find contents they are interested in. In smart TV it is more time consuming for users to select contents that they prefer compared to conventional TV. Therefore, the demand for recommendation service increases to facilitate quick and easy search for the user preferred contents.

Recommendation service automatically selects product items or information that a user might be interested in and helps him to choose things of his interest in various situations. There are two typical recommendation methods: collaborative recommendation and content-based recommendation. In general, collaborative recommendation performs better compared to content-based recommendation and it is used in diverse product domains such as movies, books, and CDs [2-4].

S. Srinivasa and V. Bhatnagar (Eds.): BDA 2012, LNCS 7678, pp. 158–167, 2012.

However, from the recommendation point of view, smart TV is significantly different in that it needs to handle a large number of contents compared to the traditional TV. Smart TV also needs to deliver services in a personalized manner, not in one-way broadcasting manner as in the traditional TV. For smart TV, the large number of users and the diversity of the user's TV viewing patterns may cause difficulties in efficient recommendation. Furthermore, in the current TV environment more than one user may share a single TV set and their view records are recorded under the unique ID of the TV set, thus it is difficult to identify individual users, which is called the hidden users problem. In this case fully personalized recommendation is difficult.

In this paper, we propose an efficient collaborative recommendation method to overcome the weaknesses of the existing methods for smart TV, particularly focusing on the points mentioned above. We also investigate integration of recommendation methods for more personalized recommendation.

This paper is organized as follows. In Section 2 we discuss some related issues and studies. In Section 3 we describe our method including an efficient scoring method for user modeling, clustering, and user-based collaborative filtering. Section 4 describes experiments using a set of real IPTV log data and performance comparison and finally Section 5 concludes the paper.

2   Related Issues and Studies

Recommendation service provides a user contents such as movies, TV programs, VODs, books and news that he might be interested in. In this case we can classify recommendation algorithms into collaborative recommendation and content-based recommendation based on information and features used in recommendation. There is also hybrid recommendation that integrates multiple recommendation methods [5-7].

Collaborative filtering (recommendation) is widely used in recommendation of diverse domains including movies, books and music. It uses only users' preference information such as log data. The collaborative filtering can be divided into user-based and item-based depending on the nature of the information used [4, 8, 9]. User-based collaboration: it recommends items that are preferred by users who are similar in preference to the target user. Item-based collaboration: it recommends items that are similar in users' preference patterns to the current item chosen and it takes into account the current item chosen in recommendation while user-based collaboration does not.

One of typical collaborative filtering methods, k-nearest neighbor method has been proposed by GroupLens. In the method the similarity between the target user and other user is represented in terms of Pearson's correlation coefficient. Based on the similarity k most similar users to the target user are selected and the preference for an item is estimated by the weighted sum of those k users' preferences for the item. The similarity p_{uv} between users u and v is shown in equation (1). The preference of user u for item c' is calculated as in equation (2). In these equations r_{uc} denotes the preference of user u for item c and $\bar{r_u}$ represents the mean of the preferences over items rated by user u. C_{uv} represents the set of items that users u and v rated and V_u represents the set of k most similar users to user u [2].

$$p_{uv} = \frac{\sum_{c \in C_{uv}} (r_{uc} - \bar{r}_u)(r_{vc} - \bar{r}_v)}{\sqrt{\sum_{c \in C_{uv}} (r_{uc} - \bar{r}_u)^2} \sqrt{\sum_{c \in C_{uv}} (r_{vc} - \bar{r}_v)^2}} \tag{1}$$

$$r_{uc'} = \bar{r}_u + \frac{\sum_{v \in V_u} p_{uv} (r_{vc'} - \bar{r}_v)}{\sum_{v \in V_u} p_{uv}} \tag{2}$$

However, the k-nearest neighbor method as memory-based suffers from the sparsity and scalability problems. The sparsity problem is that preference estimation may not be reliable when the number of items rated by a user is not sufficiently large.

[10] proposes a large volume collaborative filtering technique to solve these problems. To solve the scalability problem by clustering users, the technique selects users similar to the target user from the cluster that the target user belongs to. To solve the sparsity problem, it carries out data smoothing using the centers of clusters. This method is composed of two steps: an offline step for user clustering and data smoothing and an online step for recommendation based on the k-nearest neighbor method. This method recommends items using cluster information of users, consequently, it saves both time and resources compared to the existing methods. However, it basically uses the k-means clustering algorithm therefore it still suffers from the scalability problem in the case that the numbers of users and items are large as for smart TV.

[11] proposes ClustKNN, another method dealing with the problem of scalability After clustering, it creates a surrogate user corresponding to each cluster center and it recommends by estimating the similarity between the target user and the surrogate users. ClustKNN adopts the k-means clustering and it exhibits lower performance compared to the conventional method because it uses information of the surrogate users representing the clusters of users.

[12] evaluates the performance of a collaborative recommendation using a set of real IPTV data. In the paper, an item-based collaborative recommendation as in Amazon.com is used to recommend VOD contents and the recommendation results are analyzed in the perspectives of view rating and paid contents.

[13] proposes a TV program recommendation strategy for multiple viewers based on user profile merging. However, recommendation for multiple users is different from the hidden users problem that we address in this paper. We focus on personalized recommendation to individual users in smart TV environment where individual users are difficult to be identified, not on recommendation of TV programs to multiple users such as a family or a group of dormitory students.

3   Efficient Recommendation for Smart TV Environment

In this paper, we propose an efficient recommendation method for smart TV environment and its overall structure is shown in Figure 1 [14]. In the method, users are first clustered into groups of similar preference patterns based on their past view records, then the user-based collaborative filtering is performed and finally the result is integrated into other recommendation method to produce the more personalized recommendation results. The recommendation process is composed of 1) the user clustering step, 2) the user-based collaborative filtering step, and 3) the recommendation integration step.

3.1 User Clustering

In this step users of similar preference patterns are clustered based on their viewing records and information of selected contents. Particularly in smart TV environment users' preference patterns are so diverse that user clustering is important for more tailored services to users. Without user clustering we notice that we can hardly find any significant preference patterns that apply to all users. In the following we describe our user clustering method.

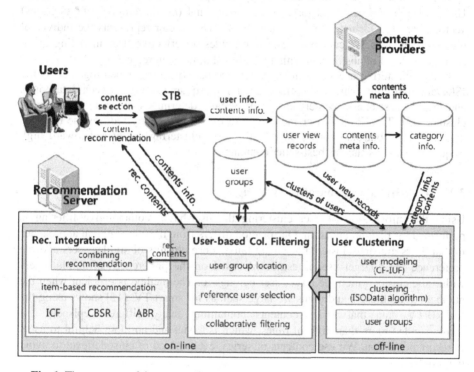

Fig. 1. The structure of the proposed recommendation method for smart TV environment

Smart TV contents are classified by genres, content providers, etc. and it constitutes the menu structure. The menu structure is built by the service provider reflecting its service policy. We analyzed a few menu structures to derive 269 common categories for smart TV contents. For user clustering each user is modeled by a frequency vector of those content categories. For smart TV, however, a simple scoring based on the frequency of choices of contents may not be appropriate because such frequencies are biased toward best-selling contents and series contents containing many episodes, and it can result a biased user clustering. In this paper we propose a new scoring technique based on CF-IUF (category frequency-inverse user frequency) as given by equations (3) and (4), a modification of TF-IDF, which is a well-known concept for information retrieval [14].

$$s_{uq} = \begin{cases} (1 + \log f_{uq}) \cdot \log \dfrac{N}{h_q}, & \text{if } f_{uq} \neq 0 \\ 0 & , \text{if } f_{uq} = 0 \end{cases} \tag{3}$$

$$F_u = \langle s_{uq_1}, s_{uq_2}, \cdots, s_{uq_m} \rangle \tag{4}$$

Here, s_{uq} represents CF-IUF of user u for category q as given in equation (3). In the equation f_{uq} denotes the number of views of category q by user u, h_q represents the number of users who viewed category q, N represents the total number of users. Using CF-IUF for each category, an m-dimensional (m: the number of categories) vector is created for user u as in equation (4). This vector represents the individual user's preferences for categories and it provides an effective user modeling in an appropriate level of abstraction with a significant dimension reduction.

In user clustering users of similar preference patterns are grouped together by the ISOData algorithm, which minimizes the sum of squared errors between data points and their closest cluster centers and automatically determines an optimal number of clusters [15]. The CF-IUF user modeling not only significantly reduces the time for user clustering but also allows an effective user clustering not biased toward high frequency contents such as best-seller contents.

3.2   User-Based Collaborative Filtering

Using the information of user clustering the user-based collaborative filtering is performed for recommendation. Our method differs from the conventional approaches in that users are first clustered to create fairly loose clusters of similar users and the preference for a target user is estimated based on the preferences of similar users within the cluster which the target user belongs to. The method is efficient when the numbers of both users and items are large such as for smart TV. [12] shows using the real IPTV data that series contents containing many different episodes are recommended more than 50% of times, which is trivial and no more helpful to users in recommendation. We use contents of series level instead of episode level to avoid the problem.

In user-based collaborative filtering we use two different correlation coefficients of Pearson's correlation coefficient (PCC) and Spearman's rank correlation coefficient (SCC) for extraction of similar users from a user cluster. PCC is a widely used similarity measure in the conventional collaborative recommendation and computed based on the number of user's views of contents as in equation (1). SCC uses the ranks of users' views and the coefficient s_{uv} between users u and v is computed as in equation (5). In the equation γ_{uc} represents the rank of preference user u for content c, and C_{uv} represents the set of contents that both of users u and v selected. Here UCFP and UCFS denote the user-based collaborative filtering using PCC and the user-based collaborative filtering using SCC, respectively. K most similar users are selected first and the preference of the target user for that content is estimated using their preferences for the target content.

$$S_{uv} = \frac{6\sum_{c \in C_{uv}}(\gamma_{uc} - \gamma_{vc})^2}{n(n^2-1)}, (n = |C_{uv}|)$$
(5)

As equation (6) shows the final preference ($\hat{r}_{uc}$) is computed as the weighted sum of preferences of similar users to the target user. In the equation V_u represents the set of similar users to user u and ω_{uv} represents the weight between users u and v.

$$\hat{r}_{uc} = \frac{\sum_{v \in V_u} \omega_{uv} r_{vc}}{\sum_{v \in V_u} \omega_{uv}}$$
(6)

We use p_{uv} in UCFP while we use s_{uv} in UCFS. The estimated preference is a real number between 0 and 1 and it represents higher preference as it approaches to 1.

3.3   Recommendation Integration

By the nature of current TV environment a user corresponds to a household unit which may contain different individual users of different preferences. Currently it is difficult to identify individual users based on user log data. Accordingly, recommendation should be household level, not individual user level. In this step we integrate different recommendation methods for more personalized recommendation. It is still difficult to evaluate the performance of personalized recommendation because individual user level log data are not available. However, we propose category match ratio as a measure for personalized recommendation. Category match ratio is a ratio of frequencies that the category of a recommended content agrees to the category of content that the user currently selected. In this paper we investigate different parallel integrations using Borda count and certainty factor. In parallel integration we simply integrate the results of two or more recommendation methods to produce the final recommendation result [5,6,7].

The Borda count method adds up the scores of individual contents recommended by different methods and the scores usually are the inverse of priority. For example, suppose that content A is recommended as the first priority and the second priority, by two different methods, respectively. Then the final score associated with content A is $1+1/2 = 1.5$.

Certainty factor is used to represent the uncertainty of knowledge in artificial intelligence. Certainty factor represents uncertainty of knowledge by a real number between -1 and 1 where 1 and -1 represent truth and falsity, respectively. Two certainty factors associated with the same knowledge are combined as in equation (7). In the equation C_a and C_b represent certainty factors of the same knowledge denoted by a and b, respectively and C_{ab} denotes combined certainty factor.

$$C_{ab} = \begin{cases} C_a + C_b - (C_a \cdot C_b), & \text{when } C_a \geq 0, C_b \geq 0 \\ C_a + C_b + (C_a \cdot C_b), & \text{when } C_a < 0, C_b < 0 \\ \dfrac{C_a + C_b}{1/\min(|C_a|, |C_b|)}, & \text{otherwise} \end{cases}$$
(7)

We use the certainty factor method to integrate different recommendation results by transforming into certainty factor in [-1, 1] preferences for each content resulted from individual recommendation methods. Certainty factors are combined according to equation (7) and the result is transformed back into the combined preference. For example, suppose that content A is given 0.9 and 0.7 as preference scores by two different recommendation methods, respectively. The two numbers are transformed into certainty factors 0.8 and 0.4, respectively. The certainty factors are combined to produce 0.88, and it is transformed back into preference score 0.94, which becomes the combined preference.

4   Experiments and Performance Comparison

We experimented with our method using a set of real IPTV log data, collected for a month (29 days) of 2009 by an IPTV service company in Korea. The data contains the total number of about 96,800,000 TV view records with about 850,000 users (TV sets) and 80,000 contents. We divided the whole data set into two parts: about 86,550,000 records of the first 26 days for training and about 10,250,000 records of the last three days for test.

We used precision, recall, success rate and coverage as performance measures. Success rate represents whether the recommended set of contents contains any content of the test set of contents the user selected. If the recommended set of contents contains any content that the user actually selected, we evaluate the recommendation as success valued 1, otherwise as failure valued 0. The final success rate is averaged over the number of recommendations. The coverage determines whether the recommended set of contents is nonempty (valued 1) or not (valued 0). The final coverage is averaged over the number of recommendations. In the experiments we evaluated the performance by assuming that a user selects one of contents in the test set and the rest is the set of contents that the user would select in the future.

We evaluated and compared the performances of UCFP and UCFS. UCFS has shown consistently better performance compared with UCFP for different numbers of user clusters and UCFP has shown better coverage than UCFS for all clusterings but lower performance in success rate, precision, and recall. The precision gets higher as the number of clusters gets smaller but the processing time gets longer. The reason is that when the number of clusters is smaller, the average size of a cluster is larger and it takes more processing time for recommendation. In this paper we experimented and compared the performance for 149 clusters as appropriate taking both processing time and performance into account.

We compared our method with other methods including the item-based collaborative recommendation as proposed in [8], the association-based recommendation as proposed in [16], and the category-based best-seller recommendation. For the association-based recommendation we used FP-tree with the minimum support of 0.1%, the minimum confidence of 30%. The category-based best-seller method recommends a fixed number of contents of the highest ranks among the contents that belong to the same category as the current content selected. In the table CBSR, ICF, and ASR denote the category-based best-seller recommendation, the item-based collaborative recommendation, and association-based recommendation, respectively. We experimented five times with randomly chosen 10,000 users and the average performances are shown in Table 1.

Table 1. Performance comparison (unit: %)

	Coverage	Success Rate	Precision	Recall
CBSR	99.88	44.33	17.7	8.72
ICF	99.72	53.41	22.14	12.18
ASR	99.27	53.1	27.49	12.42
UCFS	**96.77**	**93.58**	**77.4**	**46**
UCFP	**99.28**	**93.44**	**54.47**	**35.47**

In the case of coverage other methods except UCFS show more than 99%, which means that contents are recommended to almost all users. Even though UCFS may recommend an empty set of contents to some of users and it may affect the overall performance. Both of our proposed methods UCFP and UCFS show higher success rates compared with the conventional methods. Their precisions mean that a user actually selects at least one of the recommended contents. The precisions of UCFP and UCFS are 77.40% and 54.47%, respectively. The two methods outperform the conventional methods. We believe that user clustering causes such high precision by segmenting users into groups of users of similar preference patterns and it allows more accurate recommendation. Particularly, we notice that in precision UCFS is 20% points higher than UCFP.

Table 2. Performance comparison for recommendation integration (unit: %)

Integration Method	Rec. Method	Coverage	Suc. Rate	Precision	Recall	Cat.Match Ratio
Certainty Factor	UCFS+CBSR	99.99	92.95	44.08	33.57	20.60
	UCFS+ICF	99.98	93.00	46.65	34.65	20.40
	UCFS+ASR	99.13	73.12	54.40	38.73	13.92
	UCFP+CBSR	99.89	91.97	44.45	31.44	17.00
	UCFP+ICF	100.00	91.99	42.22	30.53	18.00
	UCFP+ASR	99.89	91.97	44.45	31.44	12.40
Borda Count	**UCFS+CBSR**	**99.99**	**93.66**	**46.38**	**34.79**	**28.20**
	UCFS+ICF	**99.98**	**93.95**	**48.50**	**36.05**	**22.60**
	UCFS+ASR	99.13	93.59	54.07	37.94	15.33
	UCFP+CBSR	**100.00**	**91.59**	**36.20**	**29.57**	**28.20**
	UCFP+ICF	**100.00**	**91.46**	**41.61**	**30.40**	**21.00**
	UCFP+ASR	99.89	92.83	46.13	33.58	13.82
Sequential Integration	UCFS-CBSR	60.06	46.70	37.26	12.21	99.90
	UCFS-CF	63.89	51.73	39.64	14.80	60.10
	UCFS-ASR	66.83	52.98	41.14	14.10	47.58
	UCFP-CBSR	65.67	41.32	26.20	10.26	99.89
	UCFP-ICF	69.27	46.37	27.62	12.70	52.26
	UCFP-ASR	79.36	49.64	29.30	12.54	30.75

We also experimented integration of different recommendation methods for more personalized and efficient recommendation. We used certainty factor and Borda count for parallel integration. The performances are compared in Table 2. We also investigated sequential integration in which the second method evaluates and recommends from among the contents recommended by the first method. The performances for different sequential integrations are shown in Table 2. As expected sequential integration improves category match ratio but it significantly reduces all other performance measures. In the table for sequential integration recommendation method X–Y denotes methods X and Y are the first and the second methods, respectively.

In summary UCFS shows the best performance among single recommendation methods. However, it should be still household level recommendation, not individual user level recommendation. For more personalized recommendation parallel integration of UCFS (or UCFP) and CBSR or ICF can be accepted as reasonable.

5   Conclusions

In this paper we propose an efficient recommendation method for smart TV, which is characterized by large numbers of users and contents. The method is composed of three steps: the user clustering step, the user-based collaborative filtering step, and the recommendation integration step. In our approach users are first clustered into groups of users of similar preference patterns. It is important to provide user-tailored services to users having so diverse preference patterns. We also propose the concept CF-IUF, a modified TF-IDF to use for user modeling. In user-based collaborative filtering we use Pearson's correlation coefficient and Spearman's rank correlation coefficient for the similarity between users to select similar users to the target user in preference patterns. Finally we investigate integration of different recommendation methods for more personalized and efficient recommendation.

We experimented to evaluate the proposed method using a set of real IPTV log data and the performances are compared with existing methods. The proposed method apparently outperforms the conventional methods and it proves a significant performance improvement for smart TV. However, for the current TV environment full personalized recommendation is difficult because of the hidden users problem. We investigated integration of different recommendation methods for more personalized and efficient recommendation. Integration of UCFS (or UCFP) and CBSR or ICF significantly improves category match ratio but it also reduces precision and recall significantly. Our future research includes deep integration of different methods for more personalized and high performance recommendation in smart TV environment.

References

1. Kim, K.H., Ahn, C.H., Hong, J.W.: Research and Standardization Trends on Smart TV. Electronics and Telecommunications Trends, 37–49 (2011)
2. Melville, P., Sindhwani, V.: Recommender Systems. In: Sammut, C., Webb, G.I. (eds.) Encyclopedia of Machine Learning, pp. 829–838. Springer US (2010)
3. Adomavicius, G., Tuzhilin, A.: Toward the Next Generation of Recommender Systems: A Survey of the State-of-the-Art and Possible Extensions. IEEE Transactions on Knowledge and Data Engineering 17(6), 734–749 (2005)

4. Kim, M.W., Kim, E.-J.: Performance Improvement in Collaborative Recommendation Using Multi-Layer Perceptron. In: King, I., Wang, J., Chan, L.-W., Wang, D. (eds.) ICONIP 2006. LNCS, vol. 4234, pp. 350–359. Springer, Heidelberg (2006)
5. Varshavsky, R., Tennenholtz, M., Karidi, R.: Hybrid Recommendation System. Google Patents (2009)
6. Claypool, M., Gokhale, A., Miranda, T., Murnikov, P., Netes, D., Sartin, M.: Combining Content-based and Collaborative Filters in an Online Newspaper. In: Proceedings of ACM SIGIR Workshop on Recommender Systems (1999)
7. Kim, B.M., Li, Q.: A Hybrid Recommendation Method based on Attributes of Items and Ratings. Journal of KIISE: Software and Applications 31(12), 1672–1683 (2004)
8. Linden, G., Smith, B., York, J.: Amazon.com Recommendations: Item-to-Item Collaborative Filtering. IEEE Internet Computing 7(1), 76–80 (2003)
9. Lowd, D., Godde, O., McLaughlin, M., Nong, S., Wang, Y., Herlocker, J.L.: Challenges and Solutions for Synthesis of Knowledge Regarding Collaborative Filtering Algorithms. Technical Reports (Electrical Engineering and Computer Science), Oregon State University (2009)
10. Xue, G.R., Lin, C., Yang, Q., Xi, W.S., Zeng, H.J., Yu, Y., Chen, Z.: Scalable collaborative filtering using cluster-based smoothing. In: Proceedings of the 28th Annual International ACM SIGIR Conference on Research and Development in Information Retrieval, pp. 114–121. ACM (2005)
11. Al Mamunur Rashid, S.K.L., Karypis, G., Riedl, J.: ClustKNN: a highly scalable hybrid model-& memory-based CF algorithm. In: Proceeding of WebKDD 2006. ACM (2006)
12. Jung, H.Y., Kim, M.S.: Collaborative Filtering Model Analysis based on IPTV Viewing Log. In: Proc. of the 37th KIISE Fall Conference, KIISE, vol. 37(1(C)), pp. 404–409 (2010)
13. Kim, E.J., Song, W.M., Song, S.Y., Kim, M.W.: An Efficient Collaborative Recommendation Technique for IPTV Services. Journal of KIISE: Software and Applications 39(5), 390–398 (2012)
14. Manning, C.D., Raghavan, P., Schutze, H.: Introduction to Information Retrieval, vol. 1. Cambridge University Press, Cambridge (2008)
15. Memarsadeghi, N., Mount, D.M., Netanyahu, N.S., Le Moigne, J., de Berg, M.: A Fast Implementation of the ISODATA Clustering Algorithm. International Journal of Computational Geometry and Applications 17(1), 71–103 (2007)
16. Lin, W., Alvarez, S.A., Ruiz, C.: Collaborative Recommendation via Adaptive Association Rule Mining. Data Mining and Knowledge Discovery 6(1), 83–105 (2002)

Materialized View Selection Using Simulated Annealing

T.V. Vijay Kumar and Santosh Kumar

School of Computer and Systems Sciences,
Jawaharlal Nehru University,
New Delhi-110067, India

Abstract. A data warehouse is designed for the purpose of answering decision making queries. These queries are usually long and exploratory in nature and have high response time, when processed against a continuously expanding data warehouse leading to delay in decision making. One way to reduce this response time is by using materialized views, which store pre-computed summarized information for answering decision queries. All views cannot be materialized due to their exponential space overhead. Further, selecting optimal subset of views is an NP-Complete problem. Alternatively, several view selection algorithms exist in literature, out of which most are empirical or based on heuristics like greedy, evolutionary etc. It has been observed that most of these view selection approaches find it infeasible to select good quality views for materialization for higher dimensional data sets. In this paper, a randomized view selection algorithm based on simulated annealing, for selecting Top-K views from amongst all possible sets of views in a multidimensional lattice, is presented. It is shown that the simulated annealing based view selection algorithm, in comparison to the better known greedy view selection algorithm, is able to select better quality views for higher dimensional data sets.

Keywords: Data Warehouse, Materialized Views, View Selection, Randomized Algorithm, Simulated Annealing.

1 Introduction

In the current scenario, the prime goal of any organization is to layout effective and efficient strategies to access and exploit data from data sources spread across the globe. As given in [42], this data can be accessed from data sources using two approaches namely the on-demand approach and the in-advance approach. In the former, the data relevant to the user query is accessed and integrated, whereas, in the latter approach, the data is accessed and integrated a prior and the query is processed against this data. The latter approach is referred to as the data warehousing approach[42] and the integrated data is stored in the central repository, referred to as a data warehouse[15]. A data warehouse stores subject oriented, integrated, time variant and non-volatile data to support decision making[15]. Unlike, traditional databases, a data warehouse stores historical or archival data, in a summarized form, accumulated over a period of time. This historical summarized data contained in a data warehouse can be used to predict future business trends. This helps in the formulation of

S. Srinivasa and V. Bhatnagar (Eds.): BDA 2012, LNCS 7678, pp. 168–179, 2012.

effective and competitive business strategies for the future. These strategies would rely on the analytical and decision queries posed on the data warehouse and their response time. Most decision queries are long and exploratory in nature. These queries are processed against a continuously growing data warehouse. Consequently, their response time is high leading to delay in decision making. This high response time needs to be reduced in order to ensure efficient decision making. Though several strategies[5, 10] have been proposed to optimize the query response time, they do not scale up with the continuously expanding data warehouse[24]. This problem to some extent has been addressed by using materialized views[26].

Materialized views, unlike traditional virtual views, store data along with definition. These contain pre-computed and summarized information, stored separately from a data warehouse, for the purpose of answering decision queries. They aim to reduce the response time of decision queries by providing answers to most of these queries without these requiring processing against a large data warehouse. This would necessitate that these views contain relevant and required information for answering decision queries. Selection of such information is referred to as a view selection problem in literature[6]. View selection deals with selecting an appropriate set of views capable of improving the query response time while conforming to constraints like storage space etc. [6, 9, 43, 44]. All possible views cannot be materialized, as they would not be able to fit within the available storage space. Further, selection of optimal subset of views is shown to be an NP-Complete problem[13]. Several alternative approaches exist for selecting materialized views of which most are empirical or heuristic based. The empirical based approaches[1, 2, 3, 4, 8, 21, 22, 23, 29, 30, 35, 41] use the past query patterns to select views for materialization. Views are also selected based on greedy [9, 11, 12, 13, 27, 31, 32, 33, 34, 36, 37, 38, 39] or evolutionary [14, 19, 45, 46] heuristics. It has been observed that most of these view selection approaches find it infeasible to select good quality views for materialization for higher dimensional data sets. As a result, the query response time is high. The query response time, in case of greedy algorithms, grows exponentially with an increase in the number of dimensions and thereby makes exhaustive optimization impracticable. Motivated by this fact, several fast randomized search heuristics [7, 16, 28] have been used to solve such problems. These approaches provide optimal execution plans for complex queries involving large numbers of aggregates. For the OLAP queries which involve 10-15 number of aggregates, resulting in 2^{10} to 2^{15} views, randomized algorithms provide a viable method for selecting subset of views for materialization [17].

Randomized algorithms consider each solution to be a state in a solution space with an associated problem specific evaluation function[16, 25]. These algorithms perform random walks on the state space via a series of moves, which are used to construct edges between connected solutions. Two solutions are considered as connected if one of them can be obtained from the other by applying one move. Algorithms in this category traverse the graph, so formed, and terminate as soon as an acceptable solution is found or when a pre-defined time limit has been exceeded. Though, algorithms for view selection exist, they do not scale up when the problem size exceeds a certain limit. Randomized algorithms have been widely used to solve such problems. These algorithms, which are based on statistical concepts, explore the large search space randomly using a problem specific evaluation function. Randomized algorithms can compute reasonably good solutions within a short period of time by achieving a trade-off between the execution time and the quality of the solution. The solution obtained by these algorithms is a near-optimal solution for complex problems having large or even unlimited search spaces. Thus, it is most appropriate for solving combinatorial problems like the view selection problem.

In this paper, a randomized view selection algorithm, based on simulated annealing [18], for selecting Top-K views, from amongst all possible sets of views in a multidimensional lattice is presented. This algorithm, referred to as simulated annealing view selection algorithm (SAVSA), considers each set of Top-K views to be a state in a solution space with an associated total cost of evaluation of all the views [13], referred to as Total View Evaluation Cost(TVEC). SAVSA is similar to iterative improvement, but it also accepts uphill moves having some probability. This probability is gradually decreased at each step leading to accepting only downhill moves in the later stages of the algorithm. This consideration of uphill moves would enable SAVSA to explore better solutions which is likelier to lead to selection of better quality Top-K views, with minimum TVEC, for materialization. SAVSA is compared with the most fundamental greedy algorithm given in [13], hereafter in this paper referred to as HRUA, on the TVEC value of the views selected by the two algorithms. SAVSA is able to select comparatively better quality views.

The paper is organized as follows: SAVSA is given in section 2. In Section 3, an example illustrating selection of views using SAVSA is given. Experiment based comparison of SAVSA and HRUA are given in section 4. Section 5 is the conclusion.

2   SAVSA

As discussed above, all views cannot be materialized due to space constraints. Also, the selection of an optimal subset of views is an NP-Complete problem [13]. Thus, there is a need for selecting a good set of views, from amongst all possible views, in a multidimensional lattice. The algorithm SAVSA uses simulated annealing, which is a randomized algorithmic technique, to select the Top-K views from a multidimensional lattice. A multidimensional lattice is briefly discussed next.

2.1   Multidimensional Lattice

Data cube in online analytical processing can be represented by a multidimensional lattice, where dimensions are organized into hierarchies. The nodes in the lattice may be views aggregated on the dimensions. As an example, consider a multidimensional lattice of views shown in Fig. 1. The index of the view is shown in parenthesis alongside the name of the view. The size of the views is shown alongside the node in the lattice. In a multidimensional lattice, direct dependencies are captured by defining an edge between views in adjacent hierarchies, i.e. edge between view A and AB, with the indirect dependencies getting captured transitively, i.e. between view A and ABC.

The view selection problem, as stated earlier, is an NP-Complete problem, as it cannot be reduced to a minimal set cover problem [20]. It is almost infeasible to find an optimal solution to such a problem because the solution space grows exponentially with increase in the problem size. One way to address this problem is by using randomized algorithms. Several randomized algorithms exist of which iterative improvement and simulated annealing are well known and widely used. The iterative improvement algorithm[16, 25], which is based on the hill climbing technique, is applied to a randomly chosen solution in the search space and continuously attempts to search its neighborhood for a better solution. In case no better solution is found, the search is terminated and the local optimum is generated as output. The limitation with this approach is that the computed local optimum is dependent on the starting solution. This limitation is overcome by simulated annealing [18] where a probability of acceptance for moving to a worse neighboring solution is also considered in order to eliminate the dependency on the starting solution. The simulated annealing technique is briefly discussed next.

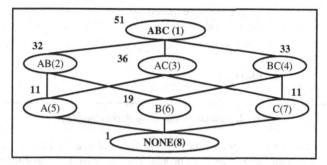

Fig. 1. 3-dimensional lattice along with size and index of each view

2.2   Simulated Annealing

Simulated annealing technique [18, 20] has its origin in statistical mechanics. The annealing of solids has been imitated to solve optimization problems. The acceptance probability is dynamically computed using the quality of the neighboring solution and the temperature value. The lower the temperature, the lesser is the likelihood of moving to a comparatively worse solution. The temperature is gradually lowered to stabilize the search and move it closer to the optimal solution. Although simulated annealing [18] follows a procedure similar to iterative improvement, it also accepts uphill moves having some probability. This probability value is decreased at every step and finally the algorithm accepts only downhill moves leading to a good local minimum. The reasons behind accepting the uphill move is that some local minima may be close to each other and are separated by a small number of uphill moves. If only downhill moves were accepted, as in the case of iterative improvement, the algorithm would stop at the first local minimum visited, missing subsequent, and possibly better, solutions. The simulated annealing algorithm is given in Fig. 2. The algorithm starts by initializing the initial state S and temperature T. There are two loops in the algorithm. The inner loop is performed for a fixed value of T, which controls the probability of uphill moves i.e. a comparatively worse solution. This probability, i.e. $e^{-\Delta C/T}$, is a monotonically increasing function of temperature T and a monotonically decreasing function of the cost difference ΔC. The inner loop executes until equilibrium is reached. The temperature T is then reduced and the inner loop executes again. The algorithm stops when the frozen condition is reached i.e. the temperature T is less than a pre-defined minimum temperature.

```
BEGIN
     Initialize S as a random state
     Initialize T as the initial temperature
     minS=S
     WHILE not (T is less than pre-defined minimum temperature) DO
          WHILE equilibrium not reached DO
               S′ = random state in neighbor(S)
               ΔC=cost(S′)-cost(S)
               IF (ΔC≤0) then S= S′
               IF (ΔC>0) then S= S′ with probability e^(−ΔC/T)
               IF cost(S) < cost(minS) then minS = S
          END DO
          T=reduce(T)
     END DO
     RETURN (minS)
END
```

Fig. 2. Simulated Annealing Algorithm [18]

The simulated annealing technique, as discussed above, has been used to select Top-K views for materialization. The proposed view selection algorithm SAVSA is discussed next.

2.3   View Selection Using SAVSA

As discussed above, it becomes almost infeasible to select views for higher dimensional data sets using deterministic algorithms as the number of views grows exponentially with increasing number of dimensions. The proposed algorithm SAVSA attempts to address this view selection problem. This algorithm is given in Fig. 3. SAVSA takes a lattice of views, with the size of each view as input. Initially, random Top-K views are chosen as the starting Top-K views V_{Top-K} with initial temperature T as T_0. SAVSA eliminates the disadvantage of hill-climbing by considering a probability for acceptance $e^{-\Delta TVEC/T}$ to decide whether to consider certain Top-K views having higher TVEC values. This acceptance probability is computed dynamically and its value depends on the Temperature T and also the difference in the TVEC value of V_{Top-K} and its neighboring Top-K views V_{Top-K}', i.e. $\Delta TVEC$. The TVEC function, defined in [40], is used to compute the cost of Top-K views.

$$TVEC\ \left(V_{Top-K}\right)= \sum_{i=1 \wedge SM_{V_i}=1}^{N} Size\ (V_i)\ +\ \sum_{i=1 \wedge SM_{V_i}=0}^{N} SizeNMA\ (V_i)$$

The inner WHILE loop works for a particular value of T computed by the outer WHILE loop. The inner loop continues until *Equilibrium* is reached i.e. a pre-specified number of iterations are completed. After the inner loop is executed, the temperature value T is reduced. The outer WHILE loop terminates when the *Frozen* condition is reached i.e. the value of T become less than pre-specified value or the $MinV_{Top-K}$ has not changed for a pre-specified number of iterations of the outer WHILE loop. The Top-K views $MinV_{Top-K}$ in the *Frozen* state is produced as output.

INPUT: Lattice of Views along with the size of each view
OUTPUT: Top- K Views $MinV_{Top-K}$
METHOD:
BEGIN
      Initial state V_{Top-K}= randomly generated Top-K views
      Initial Temperature T = T_0
      Initial Optimal State $MinV_{Top-K}=V_{Top-K}$
WHILE NOT (*Frozen*)
DO
      WHILE NOT (*Equilibrium*)
      DO
            Select a random Top-K views, say V_{Top-K}', in the neighbor of V_{Top-K}
            $\Delta TVEC= TVEC(V_{Top-K}') - TVEC(V_{Top-K})$
              IF $\Delta TVEC \leq 0$ THEN
                  Assign V_{Top-K}' to V_{Top-K}
              END IF
              IF $\Delta TVEC > 0$ THEN
                  Assign V_{Top-K}' to V_{Top-K} with probability $e^{-\Delta TVEC\ /T}$
              END IF
              IF $TVEC(V_{Top-K}) < TVEC(MinV_{Top-K})$ THEN
                  $MinV_{Top-K} = V_{Top-K}$
              END IF
      END WHILE
        T=*Reduce*(T)
END WHILE
RETURN $MinV_{Top-K}$
END

---

where

    TKV: $(V_1, V_2, \ldots, V_K)$

    TVEC : $TVEC(V_{Top-K}) = \displaystyle\sum_{i=1 \wedge SM_{V_i}=1}^{N} Size(V_i)\ +\ \sum_{i=1 \wedge SM_{V_i}=0}^{N} SizeNMA(V_i)$

where

N is the number of views in the lattice,

Size(V_i) is the size of the view V_i

SizeNMA(V_i) is size of nearest materialized ancestor of V_i

SM_{V_i} is the Status Materialized of view V_i ($SM_{V_i} =1$, if materialized, $SM_{V_i} = 0$, if not materialized)

T_0: $2\times TVEC(V_{Top-K})$

Frozen: T < pre-specified value OR $MinV_{Top-K}$ unchanged for pre-specified number of stages

Equilibrium: pre-specified number of iterations

***Reduce*(T):** pre-specified higher value (between 0 and 1) times the temperature T

Fig. 3. SAVSA

Next, an example is given that illustrates the use of the above mentioned view selection algorithm SAVSA for selecting the Top-4 views for materialization from a three dimensional lattice of views.

3   An Example

Consider the 3-dimensional lattice shown in Fig. 1. The selection of Top-4 views using SAVSA is given in Fig. 4. In the example, the inner loop, representing a stage, executes for 5 iterations for a certain value of temperature T, which is gradually reduced after each stage to 75 percent of its previous value. The acceptance

I	V_{Top-K}	$TVEC(V_{Top-K})$	V_{Top-K}'	$TVEC(V_{Top-K}')$	$\Delta TVEC$	$e^{-\Delta TVEC/T}$	V_{Top-K}	$TVEC(V_{Top-K})$	$MinV_{Top-K}$

I	V_{Top-K}	$TVEC(V_{Top-K})$	V_{Top-K}'	$TVEC(V_{Top-K}')$	$\Delta TVEC$	$e^{-\Delta TVEC/T}$	V_{Top-K}	$TVEC(V_{Top-K})$	$MinV_{Top-K}$
\multicolumn{10}{c}{V_{Top-K}= 3486, $MinV_{Top-K}$ =4386, T=520, Frozen: T < 1, Equilibrium: 5 iterations}								
1	3486	260	3426	255	-5		3426	260	3426
2	3426	255	3476	248	-7		3476	255	3476
3	3426	255	2476	240	-8		2476	248	2476
4	2476	240	2473	238	-2		2473	240	2473
5	2473	238	2873	246	8	0.9847331	2873	238	2473
\multicolumn{10}{c}{Reduce T = 0.75×T = 390}								
1	2873	246	2473	238	-8		2873	238	2473
2	2473	238	2673	243	5	0.9872613	2673	238	2473
3	2673	243	2675	237	-6		2675	238	2675
4	2675	237	2674	240	3	0.9923372	2674	237	2675
5	2674	240	3674	248	8	0.9796961	3674	237	2675
\multicolumn{10}{c}{Reduce T = 0.75×T = 292.5}								
1	3674	248	3654	245	-3		3654	237	2675
2	3654	245	7654	238	-7		7654	237	2675
3	7654	238	2654	241	3	9.90E-01	2654	237	2675
4	2654	241	2653	247	6	0.9796961	2653	237	2675
5	2653	247	2673	243	-4		2673	237	2675
\multicolumn{10}{c}{Similarly Iterations are carried out and the final two iterations are illustrated below}								
\multicolumn{10}{c}{Reduce T = 0.75×T = 1.6490303}								
1	3275	235	4275	232	-3		4275	232	4275
2	4275	232	8275	240	8	0.0078179	4275	232	4275
3	4275	232	3235	239	7	1.43E-02	4275	232	4275
4	4275	232	4273	238	6	0.0262917	4275	232	4275
5	4275	232	4273	238	6	0.0262917	4275	232	4275
\multicolumn{10}{c}{Reduce T = 0.75×T = 1.2367728}								
1	4275	232	4273	238	6	0.0078179	4275	232	4275
2	4275	232	3275	235	3	0.0884190	4275	232	4275
3	4275	232	6275	237	5	1.75E-02	4275	232	4275
4	4275	232	4235	239	7	0.0034828	4275	232	4275
5	4275	232	4875	242	10	3.08E-04	4275	232	4275
\multicolumn{10}{c}{Reduce T = 0.95×T = 0.9275796}								
\multicolumn{10}{c}{Frozen}								

Fig. 4. Selection of Top-4 views using SAVSA

probability is also considered to decide whether to choose a relatively worse view or not. This probability is high during the initial stages and thus there is high likelihood of selecting a comparatively worse view. In later stages, this probability becomes low and the search is stabilized closer to the optimal solution. The algorithm, after the frozen condition is reached, selects views 4275, i.e views BC, AB, C and A respectively, as the Top-4 views for materialization. The TVEC of these views is 232.

4   Experimental Results

Algorithms SAVSA and HRUA were implemented using JDK 1.6 in Windows-7 environment. The two algorithms were compared by conducting experiments on an Intel based 2.13 GHz PC having 3 GB RAM. The comparisons were carried out on the TVEC due to views selected by the two algorithms. Graphs were plotted to compare SAVSA and HRUA algorithms on TVEC against the number of dimensions for selecting the Top-10 views for materialization. These graphs, for the number of iterations (I) as 100, 200, 300 and 400, are shown in Fig. 5.

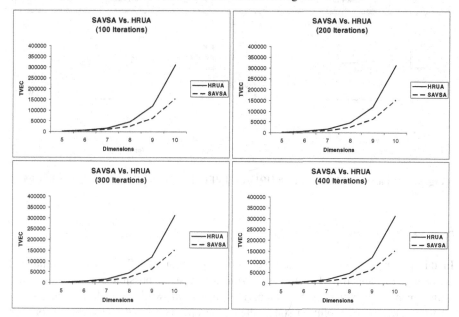

Fig. 5. Comparison of SAVSA with HRUA – TVEC Vs. Dimensions (I=100, 200, 300, 400)

It is observed from the graphs that, with increase in the number of dimensions, the TVEC value of views selected using SAVSA is lower than those selected using HRUA for each value of I.

Next, graphs for TVEC versus Top-K views for dimension 10 are plotted as shown in Fig. 6. These graphs are plotted for I=100, 200, 300, 400. The graphs show that the views selected using SAVSA have a lower TVEC than those selected using HRUA for each value of *I*. This difference is significant for higher iterations. Thus it can be inferred from the above that the Top-K views selected by SAVSA have comparatively lower TVECs than those selected using HRUA. This difference becomes significant for higher dimensional data sets and for higher values of *I*.

Further, experiments show that SAVSA is able to select Top-K views for dimensions higher than 10, whereas it becomes computationally infeasible for HRUA to select Top-K views for materialization in such cases.

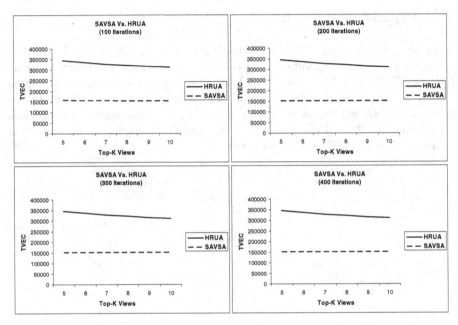

Fig. 6. Comparison of SAVSA with HRUA – TVEC Vs. Top-K Views (I=100,200,300,400)

5   Conclusion

In this paper, a randomized view selection algorithm SAVSA based on simulated annealing, for selecting Top-K views from amongst all possible sets of views in a multidimensional lattice, is presented. This algorithm, while considering each set of Top-K views to be a solution state, selects the optimal Top-K views by performing random series of downhill moves, along with some uphill moves having some probability, in the solution space. The Top-K views corresponding to local minimum, and having minimum TVEC, are then selected for materialization.   Further, experiments show that SAVSA selects comparatively better quality Top-K views that have lower TVEC when compared with those selected using the most fundamental greedy algorithm HRUA.

References

[1]   Agrawal, S., Chaudhari, S., Narasayya, V.: Automated Selection of Materialized Views and Indexes in SQL databases. In: 26th International Conference on Very Large Data Bases (VLDB 2000), Cairo, Egypt, pp. 495–505 (2000)

[2]   Aouiche, K., Jouve, P.-E., Darmont, J.: Clustering-Based Materialized View Selection in Data Warehouses. In: Manolopoulos, Y., Pokorný, J., Sellis, T.K. (eds.) ADBIS 2006. LNCS, vol. 4152, pp. 81–95. Springer, Heidelberg (2006)

[3]   Aouiche, K., Darmont, J.: Data mining-based materialized view and index selection in data warehouse. Journal of Intelligent Information Systems, 65–93 (2009)

[4]   Baralis, E., Paraboschi, S., Teniente, E.: Materialized View Selection in a Multidimansional Database. In: 23rd International Conference on Very Large Data Bases (VLDB 1997), Athens, Greece, pp. 156–165 (1997)

[5]   Chaudhuri, S., Shim, K.: Including Groupby in Query Optimization. In: Proceedings of the International Conference on Very Large Database Systems (1994)

[6]   Chirkova, R., Halevy, A.Y., Suciu, D.: A Formal Perspective on the View Selection Problem. In: Proceedings of VLDB, pp. 59–68 (2001)

[7]   Galindo-Legaria, C., Pellenkoft, A., Kersten, M.: Fast, Randomized Join-Order Selection - Why Use Transformations? In: Proc: VLDB (1994)

[8]   Golfarelli, M., Rizzi, S.: View Materialization for Nested GPSJ Queries. In: Proceedings of the International Workshop on Design and Management of Data Warehouses (DMDW 2000), Stockholm, Sweden (2000)

[9]   Gupta, H., Mumick, I.S.: Selection of Views to Materialize in a Data warehouse. IEEE Transactions on Knowledge & Data Engineering 17(1), 24–43 (2005)

[10] Gupta, A., Harinarayan, V., Quass, D.: Generalized Projections: A Powerful Approach to Aggregation. In: Proceedings of the International Conference of Very Large Database Systems (1995)

[11] Gupta, H., Harinarayan, V., Rajaraman, V., Ullman, J.: Index Selection for OLAP. In: Proceedings of the 13th International Conference on Data Engineering, ICDE 1997, Birmingham, UK (1997)

[12] Haider, M., Vijay Kumar, T.V.: Materialised Views Selection using Size and Query Frequency. International Journal of Value Chain Management (IJVCM) 5(2), 95–105 (2011)

[13] Harinarayan, V., Rajaraman, A., Ullman, J.D.: Implementing Data Cubes Efficiently. In: ACM SIGMOD, Montreal, Canada, pp. 205–216 (1996)

[14] Horng, J.T., Chang, Y.J., Liu, B.J., Kao, C.Y.: Materialized View Selection Using Genetic Algorithms in a Data warehouse System. In: Proceedings of the 1999 Congress on Evolutionary Computation, Washington D.C., USA, vol. 3 (1999)

[15] Inmon, W.H.: Building the Data Warehouse, 3rd edn. Wiley Dreamtech India Pvt. Ltd. (2003)

[16] Ioannidis, Y.E., Kang, Y.C.: Randomized Algorithms for Optimizing Large Join Queries. In: Proceedings of the 1990 ACM SIGMOD International Conference on Management of Data, ACM SIGMOD Record, vol. 19(2), pp. 312–321 (1990)

[17] Kalnis, P., Mamoulis, N., Papadias, D.: View Selection Using Randomized Search. Data and Knowledge Engineering 42(1) (2002)

[18] Kirkpatrick, S., Gelat, C., Vecchi, M.: Optimization by Simulated Annealing. Science 220, 671–680 (1983)

[19] Lawrence, M.: Multiobjective Genetic Algorithms for Materialized View Selection in OLAP Data Warehouses. In: GECCO 2006, Seattle Washington, USA, July 8-12 (2006)

[20] Lee, M., Hammer, J.: Speeding Up Materialized View Selection in Data Warehouses Using a Randomized Algorithm. Int. J. Cooperative Inf. Syst. 10(3), 327–353 (2001)

[21] Lehner, W., Ruf, T., Teschke, M.: Improving Query Response Time in Scientific Databases Using Data Aggregation. In: Proceedings of 7th International Conference and Workshop on Database and Expert Systems Applications, DEXA 1996, Zurich (1996)

[22] Lin, Z., Yang, D., Song, G., Wang, T.: User-oriented Materialized View Selection. In: The 7th IEEE International Conference on Computer and Information Technology (2007)

[23] Luo, G.: Partial Materialized Views. In: International Conference on Data Engineering (ICDE 2007), Istanbul, Turkey (April 2007)

[24] Mohania, M., Samtani, S., Roddick, J., Kambayashi, Y.: Advances and Research Directions in Data Warehousing Technology. Australian Journal of Information Systems (1998)

[25] Nahar, S., Sahni, S., Shragowitz, E.: Simulated Annealing and Combinatorial Optimization. In: Proceedings of the 23rd Design Automation Conference, pp. 293–299 (1986)

[26] Roussopoulos, N.: Materialized Views and Data Warehouse. In: 4th Workshop KRDB 1997, Athens, Greece (August 1997)

[27] Shah, B., Ramachandran, K., Raghavan, V.: A Hybrid Approach for Data Warehouse View Selection. International Journal of Data Warehousing and Mining 2(2), 1–37 (2006)

[28] Swami, A., Gupta, A.: Optimization of Large Join Queries. In: Proc. ACM SIGMOD (1988)

[29] Teschke, M., Ulbrich, A.: Using Materialized Views to Speed Up Data Warehousing. Technical Report, IMMD 6, Universität Erlangen-Nürnberg (1997)

[30] Theodoratos, D., Sellis, T.: Data Warehouse Configuration. In: Proceeding of VLDB, Athens, Greece, pp. 126–135 (1997)

[31] Valluri, S., Vadapalli, S., Karlapalem, K.: View Relevance Driven Materrialized View Selection in Data Warehousing Environment. Australian Computer Science Communications 24(2), 187–196 (2002)

[32] Vijay Kumar, T.V., Ghoshal, A.: A Reduced Lattice Greedy Algorithm for Selecting Materialized Views. In: Prasad, S.K., Routray, S., Khurana, R., Sahni, S. (eds.) ICISTM 2009. CCIS, vol. 31, pp. 6–18. Springer, Heidelberg (2009)

[33] Vijay Kumar, T.V., Haider, M., Kumar, S.: Proposing Candidate Views for Materialization. In: Prasad, S.K., Vin, H.M., Sahni, S., Jaiswal, M.P., Thipakorn, B. (eds.) ICISTM 2010. CCIS, vol. 54, pp. 89–98. Springer, Heidelberg (2010)

[34] Vijay Kumar, T.V., Haider, M.: A Query Answering Greedy Algorithm for Selecting Materialized Views. In: Pan, J.-S., Chen, S.-M., Nguyen, N.T. (eds.) ICCCI 2010, Part II. LNCS (LNAI), vol. 6422, pp. 153–162. Springer, Heidelberg (2010)

[35] Vijay Kumar, T.V., Goel, A., Jain, N.: Mining Information for Constructing Materialised Views. International Journal of Information and Communication Technology 2(4), 386–405 (2010)

[36] Vijay Kumar, T.V., Haider, M.: Greedy Views Selection Using Size and Query Frequency. In: Unnikrishnan, S., Surve, S., Bhoir, D. (eds.) ICAC3 2011. CCIS, vol. 125, pp. 11–17. Springer, Heidelberg (2011)

[37] Vijay Kumar, T.V., Haider, M., Kumar, S.: A View Recommendation Greedy Algorithm for Materialized Views Selection. In: Dua, S., Sahni, S., Goyal, D.P. (eds.) ICISTM 2011. CCIS, vol. 141, pp. 61–70. Springer, Heidelberg (2011)

[38] Vijay Kumar, T.V., Haider, M.: Selection of Views for Materialization Using Size and Query Frequency. In: Das, V.V., Thomas, G., Lumban Gaol, F. (eds.) AIM 2011. CCIS, vol. 147, pp. 150–155. Springer, Heidelberg (2011)

[39] Vijay Kumar, T.V., Haider, M.: Materialized Views Selection for Answering Queries. In: Kannan, R., Andres, F. (eds.) ICDEM 2010. LNCS, vol. 6411, pp. 44–51. Springer, Heidelberg (2012)

[40] Vijay Kumar, T.V., Kumar, S.: Materialized View Selection Using Genetic Algorithm. In: Parashar, M., Kaushik, D., Rana, O.F., Samtaney, R., Yang, Y., Zomaya, A. (eds.) IC3 2012. CCIS, vol. 306, pp. 225–237. Springer, Heidelberg (2012)

[41] Vijay Kumar, T.V., Devi, K.: Materialized View Construction in Data Warehouse for Decision Making. International Journal of Business Information Systems (IJBIS) 11(4), 379–396 (2012)

[42] Widom, J.: Research Problems in Data Warehousing. In: 4th International Conference on Information and Knowledge Management, Baltimore, Maryland, pp. 25–30 (1995)

[43] Yang, J., Karlapalem, K., Li, Q.: Algorithms for Materialized View Design in Data Warehousing Environment. The Very Large databases (VLDB) Journal, 136–145 (1997)

[44] Yousri, N.A.R., Ahmed, K.M., El-Makky, N.M.: Algorithms for Selecting Materialized Views in a Data Warehouse. In: The Proceedings of the ACS/IEEE 2005 International Conference on Computer Systems and Applications, AICCSA 2005, pp. 27–21. IEEE Computer Society (2005)

[45] Zhang, C., Yao, X., Yang, J.: Evolving Materialized Views in a Data Warehouse. In: IEEE CEC, pp. 823–829 (1999)

[46] Zhang, C., Yao, X., Yang, J.: An Evolutionary Approach to Materialized Views Selection in a Data Warehouse Environment. IEEE Transactions on Systems, Man and Cybernatics, 282–294 (2001)

Author Index